Recent Advances In
Stochastic
Operations
Research II

Recent Advances In Stochastic Operations Research II

Editors

Tadashi Dohi
Hiroshima University, Japan

Shunji Osaki
Katsushige Sawaki
Nanzan University, Japan

World Scientific

NEW JERSEY · LONDON · SINGAPORE · BEIJING · SHANGHAI · HONG KONG · TAIPEI · CHENNAI

Published by
World Scientific Publishing Co. Pte. Ltd.
5 Toh Tuck Link, Singapore 596224
USA office: 27 Warren Street, Suite 401-402, Hackensack, NJ 07601
UK office: 57 Shelton Street, Covent Garden, London WC2H 9HE

British Library Cataloguing-in-Publication Data
A catalogue record for this book is available from the British Library.

RECENT ADVANCES IN STOCHASTIC OPERATIONS RESEARCH II
Copyright © 2009 by World Scientific Publishing Co. Pte. Ltd.
All rights reserved. This book, or parts thereof, may not be reproduced in any form or by any means, electronic or mechanical, including photocopying, recording or any information storage and retrieval system now known or to be invented, without written permission from the Publisher.

For photocopying of material in this volume, please pay a copying fee through the Copyright Clearance Center, Inc., 222 Rosewood Drive, Danvers, MA 01923, USA. In this case permission to photocopy is not required from the publisher.

ISBN-13 978-981-279-166-5
ISBN-10 981-279-166-3

Printed in Singapore.

PREFACE

Operations Research uses quantitative models to analyze and predict the behavior of systems, and to provide information for decision makers. Two key concepts in Operations Research are Optimization and Uncertainty. Uncertainty is emphasized in Operations Research that could be called "Stochastic Operations Research" in which uncertainty is described by stochastic models. The typical models in Stochastic Operations Research are queueing models, inventory models, financial engineering models, reliability models, and simulation models.

International Workshop on Recent Advances in Stochastic Operations Research (2005 RASOR Canmore) was held in Canmore, Alberta, Canada, on August 25-26, 2005. At that time, a local proceedings was published and distributed to all the participants, where 40 papers were presented. After the conference, through the peer reviewing process, we published a book "Recent Advances in Stochastic Operations Research," edited by T. Dohi, S. Osaki and K. Sawaki, from World Scientific Publishing Co. Pte. Ltd., Singapore in 2007.

Following 2005 RASOR Canmore, we hosted International Workshop on Recent Advances in Stochastic Operations Research II (2007 RASOR Nanzan) at Nanzan University, Nagoya, Japan, on March 5-6, 2007. Again a local proceedings containing 43 papers was published and distributed to all the participants. After a careful peer reviewing process, this time, we are publishing a book "Recent Advances in Stochastic Operations Research II," edited by T. Dohi, S. Osaki and K. Sawaki, from World Scientific Publishing Co. Pte. Ltd., Singapore.

This conference was sponsored by the Research Center for Mathematical Sciences and Information Engineering, Nanzan University, 27 Seirei-cho, Seto-shi, Aichi 489-0863, Japan, to whom we would like to express our appreciation for their financial support. We also appreciated the financial support we received in the form of Grant-in-Aid for Scientific Research from the Ministry of Education, Sports, Science and Culture of Japan under Grant Nos. 16201035 and 18510138. Our special thanks are due to Professor Hiroyuki Okamura, Hiroshima University and Dr. Koichiro Rinsaka, Kobe Gakuin University, Japan, for their continual support from the initial planning of the conference to the final stage of editing the proceedings. Finally, we would like to thank Chelsea Chin, World Scientific Publishing Co., Singapore, for her warm help and patience.

Tadashi Dohi	Hiroshima University
Shunji Osaki	Nanzan University
Katsushige Sawaki	Nanzan University
	August 2008

LIST OF CONTRIBUTORS

I. H. Chung	— Hyundai Rotem Company, Korea
T. Dohi	— Hiroshima University, Japan
E. A. Elsayed	— Rutgers, The State University of New Jersey, USA
F. Ferreira	— University of Trás-os-Montes e Alto Douro, Portugal
T. Furuta	— Nanzan University, Japan
M. Fushimi	— Nanzan University, Japan
P. Hagmark	— Tampere University of Technology, Finland
R. Hohzaki	— National Defence Academy, Japan
Y. Ida	— Ministry of Defence, Japan
S. Inoue	— Tottori University, Japan
H. Ishii	— Osaka University, Japan
A. Ito	— Nanzan University, Japan
K. Ito	— Kinjo Gakuin University, Japan
N. Kaio	— Hiroshima Shudo University, Japan
H. Kawai	— Tottori University, Japan
H. G. Kim	— Dong-Eui University, Korea
H. Kondo	— Nanzan University, Japan
H. Kono	— Osaka University, Japan
J. Koyanagi	— Tottori University, Japan
H. Liao	— Wichita State University, USA
S. Maeji	— Kinjo Gakuin University, Japan
S. Mizutani	— Aichi University of Technology, Japan
T. Nakagawa	— Aichi Institute of Technology, Japan
T. Nakai	— Chiba University, Japan
S. Nakamura	— Kinjo Gakuin University, Japan
D. Nanba	— Tottori University, Japan
K. Naruse	— Nagoya Sangyo University, Japan
S. Osaki	— Nanzan University, Japan
A. Pacheco	— Technical University of Lisbon, Portugal
H. Ribeiro	— Polytechnic Instutute of Leiria, Portugal
K. Rinsaka	— Kobe Gakuin University, Japan
K. Sato	— Nanzan University, Japan
K. Sawaki	— Nanzan University, Japan

S. Shiode — Kobe Gakuin University, Japan
M. Tamaki — Aichi University, Japan
Y. Tamura — Hiroshima Institute of Technology, Japan
K. Tokuno — Tottori University, Japan
S. Virtanen — Tampere University of Technology, Finland
S. Yamada — Tottori University, Japan
W. Y. Yun — Pusan National University, Korea
H. Zhang — Rutgers, The State University of New Jersey, USA
Y. Zhu — Rutgers, The State University of New Jersey, USA

CONTENTS

Preface   v

List of Contributors   vii

Part A   Foundation of Stochastic Operations Research   1

A Probabilistic Proof of an Identity Related to the
Stirling Number of the First Kind   3
    M. Tamaki

A Sequential Decision Problem based on the Rate Depending on a
Markov Process   11
    T. Nakai

Search for 90/150 Cellular Automata Sequences with Maximum
Minimum-Phase-Spacing   31
    M. Fushimi, T. Furuta and A. Ito

Difference and Similarity between MONANOVA and OLS in
Conjoint Analysis   41
    H. Kono, H. Ishii and S. Shiode

Part B   Stochastic Modeling   55

A Datum Search Game and an Experimental Verification for Its
Theoretical Equilibrium   57
    R. Hohzaki and Y. Ida

An Optimal Wait Policy in Two Discrete Time Queueing Systems   69
    J. Koyanagi, D. Nanba and H. Kawai

Analysis of Finite Oscillating $GI^X/M(m)//N$ Queueing Systems   79
    F. Ferreira, A. Pacheco and H. Ribeiro

A Continuous-Time Seat Allocation Model with Up-Down Resets   99
    K. Sato and K. Sawaki

Part C   Reliability and Maintenance   115

Simulation of Reliability, Availability and Maintenance Costs   117
    P. Hagmark and S. Virtanen

Stochastic Profit Models under Repair-Limit Replacement Program   139
    T. Dohi, N. Kaio and S. Osaki

Investigation of Equivalent Step-Stress Testing Plans   151
    E. A. Elsayed, Y. Zhu, H. Zhang and H. Liao

Optimal Policy for a Two-Unit System with Two Types of Inspections   171
    S. Mizutani and T. Nakagawa

Redundancy Optimization in Multi-Level System Using Metaheuristics   183
    I. H. Chung, W. Y. Yun and H. G. Kim

Optimal Censoring Policies for the Operation of a Damage System   201
    K. Ito and T. Nakagawa

Part D   Dependable Computing   211

Optimal Sequential Checkpoint Intervals for Error Detection   213
    K. Naruse, T. Nakagawa and S. Maeji

Effective Algorithms to Estimate the Optimal Software Rejuvenation Schedule under Censoring   225
    K. Rinsaka and T. Dohi

Optimal Backup Interval of a Database System Using a
Continuous Damage Model   243
    S. Nakamura, T. Nakagawa and H. Kondo

Operational Software Performance Evaluation based on the
Number of Debuggings with Two Kinds of Restoration Scenario 253
    K. Tokuno and S. Yamada

Software Reliability Assessment with 2-Types Imperfect
Debugging Activities   271
    S. Inoue and S. Yamada

Flexible Stochastic Differential Equation Modeling for
Open-Source-Software Reliability Assessment   285
    Y. Tamura and S. Yamada

Part A Foundation of Stochastic Operations Research

A PROBABILISTIC PROOF OF AN IDENTITY RELATED TO THE STIRLING NUMBER OF THE FIRST KIND

MITSUSHI TAMAKI

Faculty of Business Administration
Aichi University
Miyoshi, Aichi, Japan
tamaki@vega.aichi-u.ac.jp

The basic assumption of the infinite formulation of the secretary problem, originally studied by Gianini and Samuels, is that, if $U_j, j = 1, 2, \ldots$, is defined as the arrival time of the jth best from an infinite sequence of rankable items, then $U_1, U_2, \ldots$, are i.i.d., uniform on the unit interval $(0,1)$. An item is referred to as a record if it is relatively best. It can be shown that a well known identity related to the Stirling number of the first kind, as given in Eq.(3) in this note, is just the identity obtained through the derivation of the probability mass function of the number of records that appear on time interval $(s,t), 0 < s < t < 1$, in two ways in the infinite formulation.

1. Introduction

A set of n rankable items (1 being the best and n the worst) appear before us one at a time in random order with all $n!$ permutations equally likely. That is, each of the successive ranks of n items constitutes a random permutation. Suppose that all that can be observed are the relative ranks of the items as they appear. If X_j denotes the relative rank of the jth item among the first j items, the sequentially observed random variables are $X_1, X_2, \ldots, X_n$. Renyi[8] has shown that

  (a) $X_1, X_2, \ldots, X_n$ are independent random variables.
  (b) $P\{X_j = i\} = 1/j, \quad 1 \leq i \leq j, 1 \leq j \leq n$.

The reader is advised to check the case $n = 3$ or 4, if he/she is not familiar to these properties of the relative ranks.

The jth item is called *candidate* if it is relatively best, i.e., $X_j = 1$ and we introduce an indicator defined as

$$I_j = \begin{cases} 1, & \text{if } X_j = 1 \\ 0, & \text{otherwise.} \end{cases}$$

Thens

$$N_n = I_1 + I_2 + \cdots + I_n \qquad (1)$$

denotes the total number of candidates. It is well known(see, e.g., Eq(2.5.9) of Arnold et al.[1] or Sec. 6.2, 6.3 and 9.5 of Blom et al.[2]) that the probability mass function of N_n is expressed as

$$p_n(k) = P\{N_n = k\} = \frac{1}{n!} \begin{bmatrix} n \\ k \end{bmatrix}, \qquad 1 \leq k \leq n,$$

where the notation $\begin{bmatrix} n \\ k \end{bmatrix}$, $1 \leq k \leq n$, $1 \leq n$ is a real number called *Stirling number of the first kind* (see an interesting paper by Knuth[7] for this notation). This number can be simply calculated from the following recursive relation

$$\begin{bmatrix} n \\ k \end{bmatrix} = \begin{bmatrix} n-1 \\ k-1 \end{bmatrix} + (n-1) \begin{bmatrix} n-1 \\ k \end{bmatrix} \qquad 1 \leq k \leq n, \ 2 \leq n$$

with $\begin{bmatrix} 1 \\ 1 \end{bmatrix} = 1$ and $\begin{bmatrix} n \\ k \end{bmatrix} = 0$ for $k = 0$ or $k > n$, or directly from

$$\begin{bmatrix} n \\ k \end{bmatrix} = (n-1)! \sum_{i_{k-1}=k-1}^{n-1} \frac{1}{i_{k-1}} \sum_{i_{k-2}=k-2}^{i_{k-1}-1} \frac{1}{i_{k-2}} \cdots \sum_{i_1=1}^{i_2-1} \frac{1}{i_1}.$$

It is noted that $\begin{bmatrix} n \\ k \end{bmatrix}$ is also interpreted as the number of permutations of n elements having k cycles(see, e.g., Graham et al.[5] or Blom et al.[2]). A typical identity of the Stirling number of the first kind is

$$\sum_{k=0}^{n} \begin{bmatrix} n \\ k \end{bmatrix} z^k = z(z+1)\cdots(z+n-1), \qquad (2)$$

which is immediate from (a), (b) and (1) if we observe that the probability generating function of the sum of the independent random variables is the product of the individual probability generating functions, i.e., $E[z^{N_n}] = \prod_{j=1}^{n} E[z^{I_j}]$. The identity with which we are concerned here is, for any positive integer k,

$$\sum_{n=k}^{\infty} \begin{bmatrix} n \\ k \end{bmatrix} \frac{z^n}{n!} = \frac{1}{k!} \left(\log \frac{1}{1-z} \right)^k, \qquad 0 < z < 1 \qquad (3)$$

as listed in Graham et al.[5](see Eq.(7.50), p.337). Multiply both sides of (2) by $v^n/n!$ ($0 < v < 1$) and then add up over n. Then

$$\sum_{n=0}^{\infty} \left(\sum_{k=0}^{n} \begin{bmatrix} n \\ k \end{bmatrix} z^k \right) \frac{v^n}{n!} = \sum_{n=0}^{\infty} \binom{z+n-1}{n} v^n = (1-v)^{-z},$$

where the last equality follows from the binomial theorem. Expanding $(1-v)^{-z} = \exp\{z \log(\frac{1}{1-v})\}$ into powers of z and comparing the coefficient of z^k on both sides yields the identity (3) with z replaced by v. Our objective is to give a probabilistic proof of this identity.

2. Probabilistic Proof

We employ the framework of the infinite secretary problem as defined and originally studied by Gianini and Samuels[3]. Let the best, second best, etc., of an infinite sequence of rankable items arrive at times $U_1, U_2, \ldots$, which are i.i.d., uniform on the unit interval $(0,1)$. For each t in this interval, let $V_i(t)$ be the arrival time of the item which is ith best among all those that arrive before time t. Then as a familiar property of random samples from a uniform distribution on $(0,1)$, we find that

$$V_i(t)\text{'s are i.i.d., uniform on } (0,t). \qquad (4)$$

An item is called *record* if it is relatively best when it appears. We denote by $N(s,t)$ the number of records that appear on time interval $(s,t), 0 < s < t < 1$, and derive the probability mass function of $N(s,t)$ in two ways.

2.1. Derivation by a forward-looking argument

One way is to relate $N(s,t)$ with a random variable defined as

$$M(s,t) = \min\{i \geq 1 : V_i(t) < s\}.$$

That is, $M(s,t)$ represents the rank of the best item that appears before s relative to all those that appear before t. Focus our attention on the arrival times of the first $m+1$ bests that appear before t. Then $M(s,t)$ takes on a value $m+1$ if and only if m bests appear after s, whereas the $(m+1)$th best appears before s. Thus we have from the property (4)

$$P\{M(s,t) = m+1\} = \frac{s}{t}\left(1 - \frac{s}{t}\right)^m, \quad m = 0, 1, 2, \ldots$$

implying that $M(s,t)$ has a geometric distribution. Conditioning on $M(s,t)$ yields

$$\begin{aligned}
P\{N(s,t) = k\} &= \sum_{m=k}^{\infty} P\{N(s,t) = k | M(s,t) = m+1\} \\
&\quad \times P\{M(s,t) = m+1\} \\
&= \sum_{m=k}^{\infty} p_m(k) P\{M(s,t) = m+1\} \\
&= \sum_{m=k}^{\infty} \frac{1}{m!} \begin{bmatrix} m \\ k \end{bmatrix} \frac{s}{t}\left(1 - \frac{s}{t}\right)^m,
\end{aligned} \tag{5}$$

where the second equality follows because, given $M(s,t) = m+1$, the arrival orders of m bests are equally likely and each of the records is identified as a candidate.

2.2. Derivation by a backward-looking argument

Another way to obtain $P\{N(s,t) = k\}$ is to trace the arrival epochs of the records backwards in time. The following lemma is crucial, which can be seen as a refinement of Theorem 1 of Gilbert and Mosteller[4](see also problem 32 of Chap.13 of Karlin and Taylor[6]).

Lemma. Let $Z_1, Z_2, \ldots$ be a sequence of random variables with Z_k uniformly distributed on time interval $(0, Z_{k-1})$, $Z_0 \equiv t < 1$. That is, $P\{Z_k \leq x \mid Z_{k-1} = a\} = x/a, 0 < x < a, k \geq 1$. If we denote by $K(s,t)$ the number of $Z_1, Z_2, \ldots$ whose values exceed s for $0 < s < t$, namely, $K(s,t) = \max\{k : Z_k > s\}$ where $\max\{\phi\} = 0$, then $K(s,t)$ is distributed as a Poisson random variable with parameter $\log(t/s)$, i.e.,

$$P\{K(s,t) = k\} = \frac{s}{t} \frac{\{\log(t/s)\}^k}{k!}. \tag{6}$$

Proof. We first show by induction on i that the distribution function of

$Z_i, i \geq 1$ is given by

$$F_i(z) = \frac{z}{t} \sum_{j=0}^{i-1} \frac{\{\log(t/z)\}^j}{j!}, \quad 0 < z < t. \tag{7}$$

The assertion (7) is true for $i = 1$, because Z_1 is uniformly distributed on time interval $(0, t)$, and so $F_1(z) = z/t$. Assume that (7) holds for i. Then since the density function of Z_i is given by

$$f_i(z) = \frac{d}{dz} F_i(z)$$
$$= \frac{\{\log(t/z)\}^{i-1}}{(i-1)!t}$$

from the induction hypothesis, $F_{i+1}(z)$ can be calculated, by conditioning on the value of Z_i, as follows.

$$F_{i+1}(z) = P\{Z_{i+1} \leq z\}$$
$$= \int_0^t P\{Z_{i+1} \leq z \mid Z_i = x\} f_i(x) dx.$$

However we easily see

$$P\{Z_{i+1} \leq z \mid Z_i = x\} = \begin{cases} 1, & \text{if } 0 < x \leq z \\ z/x, & \text{if } z < x \leq t. \end{cases}$$

Therefore

$$F_{i+1}(z) = \int_0^z 1 \cdot f_i(x) dx + \int_z^t \frac{z}{x} f_i(x) dx$$
$$= F_i(z) + \int_z^t \frac{z}{x} \frac{\{\log(t/x)\}^{i-1}}{(i-1)!t} dx$$
$$= F_i(z) + \frac{z}{t} \frac{\{\log(t/z)\}^i}{i!}$$
$$= \frac{z}{t} \sum_{j=0}^{i} \frac{\{\log(t/z)\}^j}{j!},$$

where the last equality again follows from the induction hypothesis. Thus the assertion (7) has been shown to hold for $i + 1$. We are now ready to prove (6). The event $K(s,t) = k$ occurs if and only if $Z_{k+1} \leq s < Z_k$

occurs. Thus we have from (7)

$$\begin{aligned} P\{K(s,t) = k\} &= P\{Z_{k+1} \leq s < Z_k\} \\ &= F_{k+1}(s) - F_k(s) \\ &= \frac{s}{t} \frac{\{\log(t/s)\}^k}{k!}, \end{aligned}$$

which yields (6) and the proof is complete. □

When we trace back the arrival epochs of the records starting at time t, $V_1(t)$ is interpreted as the arrival epoch of the last record, $V_1(V_1(t)) = V_1^{(2)}(t)$ as that of the second last record, $V_1(V_1^{(2)}(t)) = V_1^{(3)}(t)$ as that of the third last record and so forth. This in turn implies from (4) that $V_1^{(k)}(t)$ is distributed as Z_k in the lemma, or equivalently $N(s,t)$ is distributed as $K(s,t)$ because $N(s,t)$ is described as $N(s,t) = \max\{k : V_1^{(k)}(t) > s\}$. Thus we have from the lemma

$$P\{N(s,t) = k\} = \frac{s}{t} \frac{\{\log(t/s)\}^k}{k!}. \tag{8}$$

Putting $z = 1 - s/t$ in (5) and (8) yields the desired identity (3).

We have just shown that the identity (3) has relation with the probability mass function of the number of records that appear on some time interval in the infinite formulation of the secretary problem.

Remark. We have from the above lemma $E[K(s,t)] = \log(t/s)$. Thus, if $t = 1, s = 1/n$, then $E[K(1/n,1)] = \log n$. This result is considered as a continuous analogue of the following discrete problem: Consider a Markov chain with state space $\{0, 1, 2, \ldots\}$ and the transition probabilities

$$p_{00} = p_{10} = 1, \quad p_{ij} = \frac{1}{i}, \quad j = 0, 1, \ldots, i-1, i \geq 2,$$

where p_{ij} represents the probability that the Markov chain will, when in state i, next make a transition into state j. Let T_n denote the number of transitions needed to go from state n to state 0. Then it is well known that $E[T_n] = \sum_{j=1}^{n} 1/j$ (see, e.g., Ross[9]), implying that $E[T_n] \approx \log n$ when n is large.

Acknowledgement The author is grateful to the referee for his careful reading and helpful comments.

References

1. Arnold, B.C., Balakrishnan, N. & Nagaraja, H.N. (1998). *Records*, New York: John Wiley and Sons.
2. Blom, G., Holst, L. & Sandell, D.(1994). *Problems and snapshots from the world of probability*, New York: Springer-Verlag.
3. Gianini, J. & Samuels, S.M. (1976). The infinite secretary problem. *Annals of Probability*. 4: 418-432.
4. Gilbert, J. & Mosteller, F. (1966). Recognizing the maximum of a sequence. *Journal of American Statistical Association*. 61: 35-73.
5. Graham, R.L., Knuth, D.E. & Patashnik, O. (1989). *Concrete mathematics*. Mass.: Addison-Wesley.
6. Karlin, S. & Taylor, H.M. (1981). *A second course in stochastic processes*, Orland: Academic Press.
7. Knuth, D.E. (1992). Two notes on notation. *American Mathematical Monthly*. 99: 403-422.
8. Renyi, A. (1962). *Theorie des elements saillants d'une suite d'observations, Colloquium on Combinatorial Methods in Probability Theory, Nathematisk Institut, Aarhus University, Aarhus, Denmark*. English translation in *Selected Papers of Alfred Renyi*, Volume 2, New York: Academic Press.
9. Ross, S.M. (1997). *Introduction to probability models*, 6th ed., San Diego: Academic Press.

A SEQUENTIAL DECISION PROBLEM BASED ON THE RATE DEPENDING ON A MARKOV PROCESS*

TŌRU NAKAI

Department of Math & Informatics,
Chiba University, Yayoi-cho,
Inage-ku, Chiba 263-8522, Japan
t-nakai@facalty.chiba-u.ac.jp

It is usual to grasp an activity of the public sector as a cycle of inputs, outputs and outcomes. The inputs are the resources, the outputs are the products achieved, and the outcome is the criterion to measure the results, but it is difficult to evaluate the outcome. A sequential expenditure problem on a Markov process will be considered, and a state of this process is closely related to the outcomes. This state can be changed by expending an additional amount, and it also changes according to a Markovian transition rule based on TP_2. This stochastic order plays an important role in the Bayesian learning procedure for a partially observable Markov process. The dynamic programming formulation implies a recursive equation about the expected value obtainable under the optimal policy. There are some monotonic properties concerning this value and the optimal policy. Finally, we treat this problem on a partially observable Markov process with Bayesian learning after observing some properties under assumptions since the state can be changed by decisions. It is also possible to consider a monotonic property for this case.

1. Introduction

A concept called New Public Management for administrative management was proposed in the middle of the 1980s to activate a public sector by using management techniques in private companies and to make them more efficient. For this reason, it is necessary to establish an evaluation system for the results and a feedback system to the management cycle. It is usual to grasp the activity of the public sector as a cycle of inputs, outputs and outcomes. The inputs are the resources or expenditures, and the outputs

---

*This research was partially supported by the Grant-in-Aid for Scientific Research of the Japan Society for the Promotion of Science and Technology, and the Grant-in-Aid for Research Project of Nomura Foundation for Academic Promotion.

are the products and services achieved as a result. An outcome is a criterion to measure the performance and the results for the goal or the target. The relation among inputs and outputs is comparatively easy to evaluate when the results can be numerically measured, but, in fields where a qualitative evaluation is required, it is not easy to evaluate the performances and the results. Especially, the outcome is important to evaluate the activity of the public sector, but it is usually difficult to evaluate numerically.

In this paper, we will consider an expenditure problem in public sectors depending on outcomes. Consider an activity of public sectors like a fire service, and also consider to expend within a range of the budget for each period. For this service, there exist some relationships between the number of equipment or staffs and a rate of a population who satisfy these services. On the other hand, this rate may change as the public situation changes. Therefore, the outcome varies according to some additional expenditure from the budget, but it is also reflected by circumstances like economic conditions, environments and so on.

In order to represent these situations, we consider a stochastic process, in which the states closely relate to an outcome of a public service. By using this process, we formulate an expenditure problem as a sequential decision problem with Markovian transition. Though the outcome is an important factor to evaluate the public services, it is difficult to measure numerically. In order to treat this case, we consider a rate of a population who satisfy the services as a barometer, and this rate depends on a state of a Markov process. Since it is difficult to formulate these problems as a decision problem, it is assumed that the number of residents who satisfy the services is measured by a probability distribution on $(-\infty, \infty)$ which is considered as a state space of the process.

For this sequential expenditure problem, the problem is how much to expend to public services to improve the outcomes. A state of the process is not only reflected by public situations, but also changed by expending some additional amount within a range of the budget. In the final section, we treat this problem on a partially observable Markov process with Bayesian learning procedure, since it is not usual that the outcome is observed directly. From this respect, it is necessary to observe several relationships between prior and posterior information concerning three factors (decision, observation and transition).

In Sec. 2, an evaluation of public sectors and the relationship among inputs, outputs and outcomes are summarized according to Nakai[8]. In Nakai[10], a similar problem is analyzed for this purpose, but a model is little bit complicated, *i.e.* an outcome is treated as a state of the process. From this point, a relation between an expenditure and an outcome is not clear, and a reward function is assumed to be concave function of an outcome. In this paper, an outcome depends on a state of the process, and a reward function is defined as a function of this state. From this point, this model contains a wider class of a reward function as a function of an outcome. In Sec. 3, we summarize some properties about a case where a state of the process is observable and not reflected by other public situations as Nakai[10]. In Sec. 4, a state changes according to a Markov process, and some monotonic properties about the optimal policy and the expected value obtainable by this policy are considered under several assumptions. Finally, we will treat this problem on a partially observable Markov process with Bayesian learning procedure, where a state is not observable directly. Since a state is not only reflected by transition but also changed by decisions, a gradually condition is introduced to a probability distribution in the set of all information about unobservable state of the process. Concerning this case, a similar problem is considered in Nakai[10], but, in this paper, we will consider essential properties in detail since there exists a property investigated insufficiently.

2. Evaluation based on Outcomes and Decisions

In a private company, profit as a numerically expressed indicator can determine whether performance is good or bad. However, when the performance of the public sector is evaluated by a criterion similar to that of a private company, those in which profits cannot be taken are not sufficiently evaluated. Therefore, we are faced with the problem of how to evaluate the performance of the public sector.

Though there are various concepts concerning the management cycle of the public sector, it is usual to grasp the concept of activity of the public sector as a cycle of inputs → outputs → outcomes as Hedley[2]. In this cycle, the products or services are produced (as the outputs) based on the resources (as the inputs) expended. It is considered that the criterion or the expectation (as the target values) is achieved as outcomes of the produced things. Therefore, though it is comparatively easy to conceptualize the relation between the inputs and the outputs, it is difficult to evaluate whether it qualifies for the targets or the goals concerning the relation between the outputs and the outcomes.

In many cases, a comparatively concrete numerical value can be obtained regarding the inputs and the outputs. On the other hand, though the outcome is an important factor in evaluating the management system, measuring the outcomes and expressing them numerically is accompanied by a plethora of trouble. In order to treat this case as a sequential decision problem, we consider a rate of a population who satisfy a service as a barometer, and this rate depends on a state of a Markov process. Since we assume that the reward function depends on a state of the process, the properties of the expected reward varies according to the distribution about a number of residents. In this paper, it is assumed that a state of the process changes according to a Markov process.

3. Sequential Expenditure Problem

3.1. *Stochastic process and outcomes*

Consider an activity of public sectors like fire services, and also consider to expend within a range of the budget at each period for this service. For this service, there exist some relationships between a number of equipment or staffs and a rate of a population who satisfy this service. On the other hand, this rate may change as the public situation changes. In order to treat this case as a sequential decision problem, we consider a rate of a population who satisfy this service as a barometer, and the rate depends on a state of a Markov process. This state is not only reflected by public situations, but also changed by expending some additional amount within a range of the budget.

Consider a Markov process with state space $(-\infty, \infty)$, and, associated with each state s, a number of residents who satisfy a certain service is distributed on $(-\infty, \infty)$ with cumulative distribution function $\Phi(s)$. This means that, when a state of the Markov process is $s \in (-\infty, \infty)$, a rate of a population who satisfy a certain service is $\Phi(s)$. For a state of a Markov process, if $\Phi(s) = 1$, then all residents satisfy this service, and a number of residents who satisfy this service decreases as s decreases.

3.2. *Sequential expenditure problem*

Initially consider a sequential expenditure problem where a state s is not changed according to a stochastic process, *i.e.* a state can be changed only by expending an additional amount within a range of a budget.

When a current state is s, let x be an additional amount to expend within a range of the budget for each period, then a state changes to

$\sigma(s,x) = s(x)$ as a function of s and x. For the simplicity, we use a notation $s(x)$ instead of $\sigma(s,x)$. Let $c(x)$ be a cost to expend an additional amount x. If $c(x) = x$, a cost is equivalent to an amount of an expenditure.

Initially, we introduce a property about a function of two variables $g(x,s)$ as Ross[11] in Definition 1.

Definition 1. Whenever a function of two variables $g(x,s)$ satisfies an inequality $g(y,t) + g(x,s) \leq g(x,t) + g(y,s)$ for any x,y and s,t where $x < y$ and $s < t$, this function is called as a submodular function.

By using this definition, we introduce an assumption concerning $c(x)$ and $s(x)$. It is easy to show that a function $s(x) = 1 - e^{-x}(1-s)$ satisfies this assumption.

Assumption 1. A function of two variables $\sigma(s,x) = s(x)$ of s and x is a submodular function, i.e.

$$\sigma(t,y) - \sigma(t,x) \leq \sigma(s,y) - \sigma(s,x) \qquad (1)$$

or $t(y) - t(x) \leq s(y) - s(x)$ for any x,y and s,t where $x < y$ and $s < t$. $c(x)$ and $s(x)$ are increasing and convex functions of x, and $s(x)$ is an increasing function of s, where $c(0) = 0$ and $s(0) = s$.

If $\sigma(s,x) = s + d(x)$, then this function satisfies Eq. (1). On the other hand, even if an expenditure is the same, an improvement of a rate of a population who satisfy a certain service is different when a state of the process is not the same. By this reason, it is possible to assume $\sigma(s,x) = s + d(x)$ without loss of generality.

When there are n periods to go and a range of the budget is K, it is possible to increase a rate of a population who satisfy a certain service by expending the equipment, the staffs and so on within an amount of this range. When a state of the process is s, let $v_n(s)$ be an expected value by employing the optimal policy, then the optimality equation is

$$v_n(s) = \max_{0 \leq x \leq K} \{-c(x) + v_{n-1}(s(x))\}, \qquad (2)$$

where $v_1(s) = \max_{0 \leq x \leq K}\{-c(x) + u(s(x))\}$ with initial condition $v_0(s) = u(s)$. $u(s)$ is assumed to be an increasing and concave function of s.

By using an induction principle on n, monotonic properties are obtained concerning $v_n(s)$.

Lemma 1. $v_n(s)$ is a non-decreasing function of s, i.e. if $s \leq t$, then $v_n(s) \leq v_n(t)$. $v_n(s)$ is a non-decreasing function of n, i.e. $v_n(s) \leq v_{n+1}(s)$ for any $n \geq 0$.

When a number of remaining period is n and a state of the process is s, let $x_n^*(s)$ be an optimal amount of this problem. The following monotonic properties are obtained for the optimal policy.

Lemma 2. *Suppose that there are n periods to go and a state of the process is s, then $x_n^*(s) \leq x_n^*(t)$ for all $s \leq t$ and $n \geq 0$, and $x_{n-1}^*(s) \geq x_n^*(s)$ for any s.*

These properties are obtained for a case where a state can be changed only by expending an additional amount. In the next section, we will treat a case where a state changes according to a stochastic process in addition to expending an additional amount within a range of the budget.

4. Sequential Expenditure Problem on a Markov Process

We consider a rate of a population who satisfy a certain service or output as a barometer, and a number of residents is distributed on $(-\infty, \infty)$ which is considered as a state space. In this section, we consider a sequential expenditure problem for a case where this state changes according to a Markov process with transition probability $(p_s(t))_{-\infty \leq s \leq \infty}$, *i.e.* a state changes according to a certain stochastic process which corresponds to the uncontrollable matters in addition to expending an additional amount within a range of the budget. This is a case where, even if the equipment and the staffs grow larger, a rate, depending on a state, of a population who satisfy a certain service may decrease, *i.e.* a state can be changed by expending some additional amounts within a range of the budget and also changed by a certain stochastic process.

4.1. *Stochastic order relation*

Among random variables, consider three stochastic order relations, *i.e.* LRD(likelihood ratio order; T1), FSD(first order stochastic dominance; T2) and SSD(second order stochastic dominance; T3), according to Kijima and Ohnishi[3].

T1. Suppose that random variables X and Y have respective probability density functions $f(x)$ and $g(x)$. If $f(y)g(x) \leq f(x)g(y)$ for all x and y where $x \geq y$, then X is said to be greater than Y by means of the likelihood ratio, or simply $X \geq_{LRD} Y$.

It is easy to show that this order T1 is a partial order. Let $\mathcal{F}_{FSD} = \{u(\cdot) \mid u(x) \text{ is an increasing function of } x\}$ and $\mathcal{F}_{SSD} = \{u(\cdot) \mid$

$u(x)$ is an increasing and concave function of x}. By using these sets, two partial orders are defined as T2 and T3.

T2. Suppose random variables X and Y with density functions. If $E[u(X)] \geq E[u(Y)]$ for all $u(x)$ in $\mathcal{F}_{FSD}$, then $X \geq_{FSD} Y$.

T3. Suppose random variables X and Y with density functions. If $E[u(X)] \geq E[u(Y)]$ for all $u(x)$ in $\mathcal{F}_{SSD}$, then $X \geq_{SSD} Y$.

Among these orders, Lemma 3 is obtained as Kijima and Ohnishi[3].

Lemma 3. *Suppose two random variables X and Y. If $X \geq_{LRD} Y$, then $X \geq_{FSD} Y$, and if $X \geq_{FRD} Y$, then $X \geq_{SSD} Y$.*

4.2. Transition probability of a Markov process

For a transition probability $(p_s(t))_{-\infty \leq s \leq \infty}$ of a Markov process, let S_s be a random variable representing a state after changing according to a transition rule when the current state is s. Similarly, a random variable $S_{s(x)}$ represents a state of the process changed according to a transition rule after expending an additional amount of x when the current state is s. By using these random variables, we introduce an assumption about the transition probability of this Markov process.

Assumption 2. *For $(p_s(t))_{-\infty \leq s \leq \infty}$, if $s < s'$, then $S_{s'} \geq_{SSD} S_s$.*

If $S_{s'} \geq_{LRD} S_s$ for any s and s' where $s < s'$, then this transition probability also satisfies Assumption 2. It is easy to show Lemma 4 by T2.

Lemma 4. *If $s < s'$, then $\int_{-\infty}^{\infty} p_s(t)u(t)dt \leq \int_{-\infty}^{\infty} p_{s'}(t)u(t)dt$ for any increasing and concave function $u(s)$ of s.*

By this lemma, if $u(t)$ is an increasing and concave function of t, then $\int_{-\infty}^{\infty} p_s(t)u(t)dt$ is also an increasing function of s. If $x < y$, then $s(x) < s(y)$ by Assumption 1, and Lemma 4 implies Lemma 5, since $S_{s(y)} \geq_{SSD} S_{s(x)}$ for any x and y where $x < y$.

Lemma 5. *If $x < y$, then $\int_{-\infty}^{\infty} p_{s(x)}(t)u(t)dt \leq \int_{-\infty}^{\infty} p_{s(y)}(t)u(t)dt$ for any increasing function $u(s)$ of s.*

Similar properties are obtained if $S_{s'} \geq_{FSD} S_s$ or $S_{s'} \geq_{LRD} S_s$ for any s and s' where $s < s'$. A property $S_{s'} \geq_{LRD} S_s$ for any s and s' where $s < s'$, is equivalent to Definition 2.

Definition 2. For $\boldsymbol{P} = (p_s(t))_{s,t \in (-\infty,\infty)}$, $\begin{vmatrix} p_s(u) & p_s(v) \\ p_t(u) & p_t(v) \end{vmatrix} \geq 0$ for any s, t, u and v, where $s \leq t$ and $u \leq v$ ($s, t, u, v \in (-\infty, \infty)$).

If $\boldsymbol{P} = (p_s(t))_{s,t \in (-\infty,\infty)}$ satisfies this property, then $\boldsymbol{P}$ is said to be total positive of order two, or simply TP_2. The property called TP_2 plays an important role in the Bayesian learning procedure for a partially observable Markov process such as a sequential assignment problem, a job search problem and so on as Nakai [9].

Since $S_{s(y)} \geq_{LRD} S_{s(x)}$ for an increasing function $s(x)$ where $x \leq y$, an inequality $S_{s(y)} \geq_{LRD} S_{s(x)}$ is represented as follows.

Lemma 6. Suppose $\boldsymbol{P} = (p_s(t))_{s,t \in (-\infty,\infty)}$ and an increasing function $s(x)$ of x. If $x \leq y$ and $u \leq v$ ($u, v \in (-\infty, \infty)$) for any s, t, u and v, then $\begin{vmatrix} p_{s(x)}(u) & p_{s(x)}(v) \\ p_{s(y)}(u) & p_{s(y)}(v) \end{vmatrix} \geq 0$ for any s ($s \in (-\infty, \infty)$).

4.3. Sequential decision model

Suppose there are n periods to go and a range of the budget is K. When a state of the process is s, let $V_n(s)$ be the expected reward obtainable under the optimal policy, then the principle of the optimality implies

$$V_n(s) = \max_{0 \leq x \leq K} \left\{ -c(x) + \int_{-\infty}^{\infty} p_{s(x)}(t) V_{n-1}(t) dt \right\}, \quad (3)$$

with $V_0(s) = \int_{-\infty}^{\infty} p_{s(x)}(t) u(t) dt$, and $s(x)$ is a state after expending an additional amount x when a current state is s. This $V_n(s)$ has a monotonic property with respect to s as Lemma 7. It is also possible to show monotonic properties about the optimal policy by a method similar to one used in Nakai[10] under Assumption 3.

Lemma 7. $V_n(s)$ is a non-decreasing function of s, i.e. if $s < s'$, then $V_n(s) \geq V_n(s')$.

Assumption 3. For $(p_s(t))_{-\infty \leq s \leq \infty}$, $\int_{-\infty}^{\infty} p_s(t) u(t) dt$ is concave with respect to s for an increasing and concave function $u(t)$ of t, and

$\int_{-\infty}^{\infty} p_s(t)u(t)dt - u(s)$ is decreasing with respect to s for an increasing and concave function $u(t)$ of t.

Let $(p_s(t))_{s,t\in(-\infty,\infty)}$ be a transition probability of a Markov process with state space $(-\infty, \infty)$, and $p_s(t) = \dfrac{1}{\sqrt{2\pi}\sigma} e^{-\dfrac{(t-s)^2}{2\sigma^2}}$. By simple calculations, this transition probability satisfies this assumption, and $p_s(u)p_t(v) \geq p_t(u)p_s(v)$ for any s,t,u,v where $s \leq t$ and $u \leq v$, i.e. this transition probability is TP$_2$. Nakai[10] shows another example when the state space is $[0,1]$.

Proposition 1. *When there are n periods to go and a state of a process is s, let the optimal amount of the expenditure be $x_n^*(s)$, then $x_n^*(s) \leq x_n^*(s')$ for all $s \leq s'$ $(n \geq 0)$, and $x_{n-1}^*(s) \geq x_n^*(s)$ for all $n \geq 1$ $(s \in (-\infty, \infty))$.*

In Sec. 3, Lemma 1 shows a monotonic property of $v_n(s)$ concerning n of a number of remaining periods. Finally, we consider a monotonic property of $V_n(s)$ for n. Unlike to usual sequential decision problems, it is not possible to quit the public service, even if the expected reward will become worse. Therefore, there exist two cases where $V_n(s)$ is non-increasing or non-decreasing with respect to n, according to a reward function $u(s)$ and a transition probability. If we assume $V_{n-1}(s) \leq V_{n-2}(s)$ for any s, then $V_n(s) \leq V_{n-1}(s)$ since

$$V_n(s) = \max_{0 \leq x \leq K} \left\{ -c(x) + \int_{-\infty}^{\infty} p_{s(x)}(t)V_{n-1}(t)dt \right\}$$

$$V_{n-1}(s) = \max_{0 \leq x \leq K} \left\{ -c(x) + \int_{-\infty}^{\infty} p_{s(x)}(t)V_{n-2}(t)dt \right\}$$

and $\int_{-\infty}^{\infty} p_{s(x)}(t)V_{n-1}(t)dt \leq \int_{-\infty}^{\infty} p_{s(x)}(t)V_{n-2}(t)dt$ by the induction assumption. On the other hand, if we assume an inequality $V_{n-1}(s) \geq V_{n-2}(s)$ for any s, then $V_n(s) \geq V_{n-1}(s)$. By the induction principle on n, the monotonic properties of $V_n(s)$ concerning n is determined by a property of $V_1(s)$. Since $V_1(s) = \max_{0 \leq x \leq K} \left\{ -c(x) + \int_{-\infty}^{\infty} p_{s(x)}(t)u(t)dt \right\}$ and $V_0(s) = u(s)$, if $V_1(s) \geq V_0(s)$, then $V_n(s)$ is a non-decreasing function of n, and, otherwise, $V_n(s)$ is a non-increasing function of n.

Let $u(s)$ be a convex function of s. For a random variable $S_{s(x)}$ representing a new state of the process by expending an additional amount x

when a current state is s, if $E[S_{s(0)}] \geq s$, then $\int_{-\infty}^{\infty} p_s(t)u(t)dt \geq u(s)$ by the Jensen's inequality. By this inequality,

$$V_1(s) \geq -c(0) + \int_{-\infty}^{\infty} p_{s(0)}(t)u(t)dt = \int_{-\infty}^{\infty} p_s(t)u(t)dt \geq u(s) = V_0(s),$$

i.e. $V_1(s) \geq V_0(s)$, and, therefore, $V_n(s)$ is a non-increasing function of n. This is a case where the expected reward is greater than the expected reward without expending additional amount.

5. Partially Observable Markov Process and Learning Procedure

5.1. *Partially observable Markov process and information*

Consider a Markov process with state space $(-\infty, \infty)$ and transition probability $(p_s(t))_{s,t \in (-\infty,\infty)}$, then $\boldsymbol{p}_s = (p_s(t))_{t \in (-\infty,\infty)}$ is a probability distribution on $(-\infty, \infty)$ for any $s \in (-\infty, \infty)$. In subsequent sections, the state of this process can not be observed, i.e. this sequential decision problem will be treated on a partially observable Markov process, and, therefore, the property T1 is assumed for the transition probability $(p_s(t))_{s,t \in (-\infty,\infty)}$ as Assumption 4 instead of Assumption 2.

Assumption 4. *For* $(p_s(t))_{-\infty \leq s \leq \infty}$, *if* $s < s'$, *then* $S_{s'} \geq_{LRD} S_s$.

Information about unobservable state is assumed to be a probability distribution μ on the state space $(-\infty, \infty)$. Let $\mathcal{S}$ be a set of all information about unobservable state, then

$$\mathcal{S} = \left\{ \mu = (\mu(s))_{s \in (-\infty,\infty)} \,\middle|\, \int_{-\infty}^{\infty} \mu(s)ds = 1, \mu(s) \geq 0 \, (s \in (-\infty,\infty)) \right\}.$$

Among informations in $\mathcal{S}$, we introduce an order by using a property T1, i.e. for two probability distributions μ, ν on $(-\infty, \infty)$, if $\mu(s')\nu(s) \leq \mu(s)\nu(s')$ for any s, s' $(s \leq s', s, s' \in (-\infty, \infty))$ and $\mu(s')\nu(s) < \mu(s)\nu(s')$ at least one pair of s and s', then μ is said to be greater than ν, or simply $\mu \succ \nu$. This order is a partial order and also said to be TP_2. Under this definition, when $\mu \succeq \nu$ $(\mu, \nu \in \mathcal{S})$, the ratio $\dfrac{\mu(s)}{\nu(s)}$ of the densities increases whenever $\nu(s) \neq 0$ as s becomes large. Concerning this order relation, Lemma 8 is also obtained from Lemma 3. In this lemma, $F_\mu(x) = \int_{-\infty}^{\infty} \mu(s) F_s(x) ds$ is a weighted distribution function as De Vylder[1].

Lemma 8. *If $\mu \succeq \nu$ in $\mathcal{S}$, then $\int_{-\infty}^{\infty} h(x)dF_{\mu}(x) \geq \int_{-\infty}^{\infty} h(x)dF_{\nu}(x)$ for a non-decreasing non-negative function $h(x)$ of x.*

For prior information μ, let $\overline{\mu}(s)$ be posterior distribution on the state space after moving forward by one unit of time by making a transition to a new state according to a transition probability, then

$$\overline{\mu}(s) = \int_{-\infty}^{\infty} \mu(t) p_t(s) dt. \qquad (4)$$

For this $\overline{\mu} = (\overline{\mu}(t))_{t \in (-\infty, \infty)}$, Lemma 9 is obtained as Nakai[7] and others.

Lemma 9. *If $\mu \succ \nu$, then $\overline{\mu} \succ \overline{\nu}$.*

5.2. *Learning procedure*

Associated to each state s ($s \in (-\infty, \infty)$), there exists a non-negative random variable Y_s as an information process, *i.e.* an observation process exists to obtain information about unobservable state. We introduce Assumption 5 concerning the random variables Y_s ($s \in (-\infty, \infty)$), since we employ the Bayes' theorem as a learning procedure. For each state s, the random variables Y_s are absolutely continuous with density $f_s(y)$ ($s \in (-\infty, \infty)$).

Assumption 5. *For random variables $\{Y_s\}_{s \in (-\infty, \infty)}$, if $s \leq s'$, then $Y_{s'} \succeq Y_s$ ($s, s' \in (-\infty, \infty)$), i.e. Y_s is increasing with respect to s by means of the likelihood ratio.*

In Assumption 5, $Y_s \succeq Y_{s'}$ implies that; if $x < y$, then $f_s(y) f_{s'}(x) \leq f_s(x) f_{s'}(y)$ for s and s' where $s \leq s'$ ($s, s' \in (-\infty, \infty)$). From this fact, the random variable Y_s takes on smaller values as s becomes smaller. The property T1 concerning transition probability implies that the probability of moving from the current state to 'better' states increases with improvement in the current state. From this fact, as a number s associated with each state becomes larger, the probability to make a transition into the higher class increases.

Regarding unobservable state of the process, we improve information by observing $\{Y_s\}_{s \in (-\infty, \infty)}$. When prior information is μ, we first observe these random variables $\{Y_s\}_{s \in (-\infty, \infty)}$ and improve information about it by employing the Bayes' theorem. After that, we see time moving forward by one unit and thus this process will make a transition to a new state. It is also possible to formulate and analyze this model by other order. In this subsection, we only consider a relation between learning and transition, and a relation to an expenditure will be treated in the next subsection. If an observation is y, we improve information as $\boldsymbol{\mu}(y) = (\mu(y)(s))_{s\in(-\infty,\infty)} \in \mathcal{S}$ by employing the Bayes' theorem where

$$\mu(y)(s) = \frac{\mu(s)f_s(y)}{\displaystyle\int_{-\infty}^{\infty} \mu(s)f_s(y)ds}. \tag{5}$$

After making a transition into a new state according to $\boldsymbol{P}$, information at the next stage becomes $\overline{\boldsymbol{\mu}(y)} = (\overline{\mu(y)}(s))$ as

$$\overline{\mu(y)}(s) = \int_{-\infty}^{\infty} \mu(y)(t)p_t(s)dt. \tag{6}$$

Regarding a relationship between prior information $\boldsymbol{\mu}$ and posterior information $\overline{\boldsymbol{\mu}(x)}$, Lemma 10 is obtained under Assumptions 4 and 5, which is obtained in Nakai[7] and others.

Lemma 10. *If $\boldsymbol{\mu} \succ \boldsymbol{\nu}$, then $\boldsymbol{\mu}(y) \succ \boldsymbol{\nu}(y)$ and $\overline{\boldsymbol{\mu}(y)} \succ \overline{\boldsymbol{\nu}(y)}$ for all y. For any $\boldsymbol{\mu}$, $\boldsymbol{\mu}(y)$ and $\overline{\boldsymbol{\mu}(y)}$ are increasing functions of y, i.e. $\boldsymbol{\mu}(y) \succ \boldsymbol{\mu}(y')$ and $\overline{\boldsymbol{\mu}(y)} \succ \overline{\boldsymbol{\mu}(y')}$ where $y > y'$.*

Lemma 10 implies that an order relation among prior information $\boldsymbol{\mu}$ is preserved in $\boldsymbol{\mu}(y)$ and posterior information $\overline{\boldsymbol{\mu}(y)}$. Furthermore, for same prior information $\boldsymbol{\mu}$, as y increases, posterior information $\overline{\boldsymbol{\mu}(y)}$ becomes better by means of the likelihood ratio.

It is possible to generalize these discussions as Nakai[7], and also to apply this partially observable Markov process to sequential decision problems as Nakai[4,5,6] and so on. On the contrary to these works, as for an optimal expenditure problem treated here, the state changes according to a transition rule of the Markov process in addition to an additional expenditure (decision), i.e., it is not only reflected by the transition rule, but also changed by decisions. From this respect, we introduce a gradually condition in Sec. 5.3 and observe several relationships between prior and posterior information concerning three factors (decision, observation and transition) in Sec. 5.4.

5.3. *Gradually condition*

To observe some properties about a sequential expenditure problem with incomplete state information, a condition concerning probability distributions on the state space will be induced in this section. In the subsequent sections, we treat a case $\sigma(s,x) = s + d(x)$ without loss of generality as

A Sequential Decision Problem Based on the Rate Depending on a Markov Process

stated in Sec. 3.2. When prior information about unobservable state of the process is μ, let $\mu_x = (\mu_x(s))$ be a probability distribution on the state space after expending an additional amount x, i.e. $\mu_x(s) = \mu(s - d(x))$. For prior information μ, let $\overline{\mu_x} = (\overline{\mu_x}(s))$ be porterior distribution on the state space by making a transition to a new state according to a transition rule after expending an additional amount x as

$$\overline{\mu_x}(s) = \int_{-\infty}^{\infty} \mu_x(t) p_t(s) dt = \int_{-\infty}^{\infty} \mu(t) p_{t(x)}(s) dt. \qquad (7)$$

Since $s(0) = s$, $\overline{\mu} = \overline{\mu}_0$.

Definition 3. For a probability distribution μ in $\mathcal{S}$, if $\dfrac{\mu(s)}{\mu(s')} \geq \dfrac{\mu(t)}{\mu(t')}$ for any $s < s'$, $t < t'$ where $s < t, s' < t'$ and $s - s' = t - t' = c < 0$, then we say that μ satisfies a *gradually condition*.

If μ satisfies a gradually condition, then it is easy to show that μ_x also satisfies a gradually condition.

Example 1. For a normal distribution on the state space where $\mu(s) = \dfrac{1}{\sqrt{2\pi}\sigma} e^{-\frac{(s-a)^2}{2\sigma^2}}$, we have $\dfrac{\mu(s)}{\mu(s')} = e^{\frac{c(s'+s-2a)}{2\sigma^2}}$. Since $c < 0$, $c(s' + s - 2a)$ is non-increasing with respect to $s' + s$, and, therefore, this μ satisfies a gradually condition.

For any information μ in $\mathcal{S}$, $\overline{\mu}(t)$ is posterior distribution on the state space after making a transition to a new state as Eq. (4). In order to show the next property, Assumption 6 is induced for the transition probability.

Assumption 6. *For any* $s < s'$, $t \leq t'$ *and* $u < v$, $p_u(s)p_v(t') - p_u(t)p_v(s') \geq p_v(s)p_u(t') - p_v(t)p_u(s')$, *i.e.*, $\begin{vmatrix} p_u(s) & p_u(t) \\ p_v(s') & p_v(t') \end{vmatrix} \geq \begin{vmatrix} p_v(s) & p_v(t) \\ p_u(s') & p_u(t') \end{vmatrix}.$

Lemma 11. *If μ satisfies a gradually condition, then $\overline{\mu}$ also satisfies a gradually condition.*

Proof. For any $s < t$ and $s' < t'$ where $s - s' = t - t' = c < 0$, it is sufficient to show $\dfrac{\overline{\mu}(s)}{\overline{\mu}(s')} \geq \dfrac{\overline{\mu}(t)}{\overline{\mu}(t')}$. By simple calculations, Assumption 6 implies the inequality

$$\int_{-\infty}^{\infty} \mu(u)p_u(s)du \int_{-\infty}^{\infty} \mu(v)p_v(t')dv \geq \int_{-\infty}^{\infty} \mu(u)p_u(t)du \int_{-\infty}^{\infty} \mu(v)p_v(s')dv,$$

and this completes the proof. $\square$

Lemma 12. *If μ satisfies a gradually condition, then $\overline{\mu_x}$ satisfies a gradually condition.*

Proof. Since μ satisfies a gradually condition, μ_x also satisfies a gradually condition, and, therefore, $\overline{\mu_x}$ satisfies a gradually condition by Lemma 11. □

Example 2. A normal distribution $p_s(t) = \dfrac{1}{\sqrt{2\pi}\sigma} e^{-\frac{(t-s)^2}{2\sigma^2}}$ distributed on the state space satisfies Assumption 6 by simple calculations.

Next consider Assumption 7 to investigate a gradually condition about posterior information $\overline{\mu(y)}$.

Assumption 7. *The distribution function $f_s(y)$ of a random variable Y_s ($s \in (-\infty, \infty)$) satisfies $\dfrac{f_s(y)}{f_{s'}(y)} \geq \dfrac{f_t(y)}{f_{t'}(y)}$ for any $s < s'$ and $t < t'$ where $t - s = t' - s' > 0$.*

When an observation y is obtained from information process, posterior information $\overline{\mu(y)} = (\overline{\mu(y)(s)})$ as Eq. (6) has a following property.

Lemma 13. *If μ satisfies a gradually condition, then $\mu(y)$ also satisfies a gradually condition for any y.*

Proof. Note that the denominators are the same for any observation y. Since prior information μ satisfies a gradually condition, we have $\mu(s)/\mu(s') \geq \mu(t)/\mu(t')$. Assumption 7 implies $f_s(y)/f_{s'}(y) \geq f_t(y)/f_{t'}(y)$. Combining these two inequalities yield $\dfrac{\mu(s)f_s(y)}{\mu(s')f_{s'}(y)} \geq \dfrac{\mu(t)f_t(y)}{\mu(t')f_{t'}(y)}$. This inequality implies $\dfrac{\mu(y)(s)}{\mu(y)(s')} \geq \dfrac{\mu(y)(t)}{\mu(y)(t')}$ for any $s < s'$, $t < t'$ where $s - s' = t - t' = c < 0$. □

Lemma 14. *If μ satisfies a gradually condition, then $\overline{\mu(y)}$ also satisfies a gradually condition for any y.*

Proof. Since prior information μ satisfies a gradually condition, Lemma 13 yields that $\mu(y)$ also satisfies a gradually condition for any y. By Lemma 11, $\overline{\mu(y)}$ also satisfies a gradually condition for any y, and this implies Lemma 14. □

Example 3. For a normal distribution $f_s(y) = \dfrac{1}{\sqrt{2\pi}\sigma}e^{-\frac{(y-s)^2}{2\sigma^2}}$, it is easy to show $\dfrac{f_s(s)}{f_{s'}(s')} > \dfrac{f_t(y)}{f_{t'}(y)}$ for any $s < s'$, $t < t'$ where $s < t$ and $t+t' > s+s'$, i.e. these distributions satisfy Assumption 7.

5.4. Monotonic property

Concerning a probability distribution μ in $\mathcal{S}$, the notations used in this section are summarized as follows when prior probability distribution on the state space is μ.

μ: a probability distribution as prior information
$\overline{\mu}$: a probability distribution after making a transition to a new state as Eq. (4)
μ_x: a probability distribution after expending an additional amount x
$\mu(y)$: a probability distribution improved by using an observation y according to the Bayes' theorem as Eq. (5)
$\overline{\mu(y)}$: a probability distribution after changing to a new state according to P as Eq. (6) when prior information is $\mu(y)$
$\overline{\mu_x}$: a probability distribution after changing to a new state according to P as Eq. (8) with expending an additional amount x when prior information is μ
$\overline{\mu(y)_x}$: a probability distribution after changing to a new state according to P as Eq. (9) with expending an additional amount x when prior information is $\mu(y)$

When prior information about unobservable state space is μ, let

$$\overline{\mu_x}(s) = \int_{-\infty}^{\infty} \mu(t) p_{t(x)}(s) dt \qquad (8)$$

be a probability distribution on the state space after expending an additional amount x.

In this paper, an order of learning, expending and transition is considered as follows, i.e. when prior information is μ, first observe a realized value y from information process and improve information about it as $\mu(y)$ by using the Bayes' theorem, and after expending an additional amount x, we see time moving forward by one unit according to P by making a transition to a new state. Information about this new state becomes

$\overline{\mu(y)_x} = (\overline{\mu(y)_x}(s))$ where

$$\overline{\mu(y)_x}(s) = \int_{-\infty}^{\infty} \mu(y)(t) p_{t(x)}(s) dt \qquad (9)$$

Initially, we show two monotonic properties about μ_x for any amount x of an expenditure.

Lemma 15. *When μ satisfies a gradually condition, if $x > x'$, then $\mu_x \succeq \mu_{x'}$.*

Proof. It is sufficient to show $\mu_x(t')\mu_{x'}(t) \geq \mu_x(t)\mu_{x'}(t')$ for any $t \leq t'$, i.e. $\mu(t' - d(x))\mu(t - d(x')) \geq \mu(t - d(x))\mu(t' - d(x'))$ for any $t \leq t'$, since $\mu_x = (\mu_x(t))$ and $\mu_{x'} = (\mu_{x'}(t))$. This inequality is equivalent to $\dfrac{\mu(t' - d(x))}{\mu(t' - d(x'))} \geq \dfrac{\mu(t - d(x))}{\mu(t - d(x'))}$. Because $t' - d(x) < t' - d(x')$ and $t - d(x) < t - d(x')$,

$$(t' - d(x)) - (t' - d(x')) = (t - d(x)) - (t - d(x')) = d(x') - d(x) < 0.$$

Since μ satisfies a gradually condition, $\dfrac{\mu(s)}{\mu(s')} \geq \dfrac{\mu(t)}{\mu(t')}$ for any $s < s'$ and $t < t'$ where $s - s' = t - t' = c < 0$, and this implies Lemma 15. □

Lemma 16. *When μ and ν satisfy a gradually condition, if $\mu \succeq \nu$, then $\mu_x \succeq \nu_x$ for any $x(\geq 0)$.*

Proof. It is sufficient to show $\mu_x(t')\nu_x(t) \geq \mu_x(t)\nu_x(t')$ for any $t \leq t'$, i.e.

$$\mu(t' - d(x))\nu(t - d(x)) \geq \mu(t - d(x))\nu(t' - d(x))$$

for any $t \leq t'$ since $\mu_x = (\mu_x(t))$ and $\nu_x = (\nu_x(t))$. Since $\mu \succeq \nu$ and $t - d(x) < t' - d(x)$. $\dfrac{\mu(t' - d(x))}{\nu(t' - d(x))} \geq \dfrac{\mu(t - d(x))}{\nu(t - d(x))}$ for any $t \leq t'$. This implies this lemma. □

Combining the properties of this section yields following lemmas concerning monotonic properties about posterior distribution.

Lemma 17. *When μ and ν satisfy a gradually condition, if $\mu \succeq \nu$, then $\overline{\mu_x} \succeq \overline{\nu_x}$ and $\overline{\mu(y)_x} \succeq \overline{\nu(y)_x}$ for any $x(\geq 0)$.*

Proof. Since μ and ν satisfy a gradually condition, $\mu \succeq \nu$ implies $\mu_x \succeq \nu_x$ for any amount $x(\geq 0)$ by Lemma 16. Lemma 9 implies $\overline{\mu_x} \succeq \overline{\nu_x}$ for any

A Sequential Decision Problem Based on the Rate Depending on a Markov Process 27

$x(\geq 0)$. On the other hand, Lemma 10 yields $\boldsymbol{\mu}(y) \succeq \boldsymbol{\nu}(y)$ and Lemma 13 implies that $\boldsymbol{\mu}(y)$ and $\boldsymbol{\nu}(y)$ satisfy a gradually condition. Lemma 16 implies $\boldsymbol{\mu}(y)_x \succeq \boldsymbol{\nu}(y)_x$, and, therefore, $\overline{\boldsymbol{\mu}(y)_x} \succeq \overline{\boldsymbol{\nu}(y)_x}$ for any $x(\geq 0)$ since $\boldsymbol{\mu}(y)_x$ and $\boldsymbol{\nu}(y)_x$ satisfy a gradually condition. □

Lemma 18. *When $\boldsymbol{\mu}$ satisfies a gradually condition, if $y > y'$, then* $\overline{\boldsymbol{\mu}(y)_x} \succeq \overline{\boldsymbol{\mu}(y')_x}$.

Proof. Since $\boldsymbol{\mu}$ satisfies a gradually condition, if $y > y'$, then $\boldsymbol{\mu}(y) \succeq \boldsymbol{\mu}(y')$ by Lemma 10. Therefore, Lemma 17 implies $\overline{\boldsymbol{\mu}(y)_x} \succeq \overline{\boldsymbol{\mu}(y')_x}$ for any amount $x(\geq 0)$ since $\boldsymbol{\mu}(y)$ and $\boldsymbol{\mu}(y')$ satisfy a gradually condition by Lemma 13. □

Lemma 19. *When $\boldsymbol{\mu}$ satisfies a gradually condition, if $x > x'$ for any amount of the expenditure, then* $\overline{\boldsymbol{\mu}(y)_x} \succeq \overline{\boldsymbol{\mu}(y)_{x'}}$.

Proof. Since $\boldsymbol{\mu}$ satisfies a gradually condition, $\boldsymbol{\mu}(y)$ also satisfies a gradually condition by Lemma 14, and, therefore, Lemma 15 implies $\boldsymbol{\mu}(y)_x \succeq \boldsymbol{\mu}(y)_{x'}$ for any observation y since $x > x'$. By Lemma 16, we have $\overline{\boldsymbol{\mu}(y)_x} \succeq \overline{\boldsymbol{\mu}(y)_{x'}}$ for any observation y since $\boldsymbol{\mu}(y)_x$ and $\boldsymbol{\mu}(y)_{x'}$ satisfy a gradually condition. □

5.5. Sequential expenditure problem: an incomplete information case

In this section, the current state of the process is not known directly, and information about this state is obtained through information process as Sec. 5.2. This problem is formulated by using a partially observable Markov process treated in Sec. 5.1.

In Sec. 2, we consider a rate of a population who satisfy a service as a barometer of an outcome, and this rate depends on a state of a Markov process. But it is not usual that the outcome is observed directly. From this reason, we will treat this problem on a partially observable Markov process, where the state is not observed directly, *i.e.*, the decision maker obtains some information about unobservable outcome by observing a realized value obtained from the information process Y_s.

Consider a sequential expenditure problem where the state changes according to a partially observable Markov process treated in Sec. 5.1. Associated to each state s ($s \in (-\infty, \infty)$), there exists a random variable Y_s as an observation process, and information regarding an unobservable state is obtained through this process. After observing these Y's, we improve information by employing the Bayes' theorem as a learning procedure. When prior information is μ, first observe a realized value y from information process and improve information about it as $\mu(y)$ by using the Bayes' theorem. After expending an additional amount x within a range of a budget, posterior information becomes $\mu(y)_x$, and we see time moving forward by one unit according to $(p_{s(x)}(t))_{-\infty \leq s \leq \infty}$ whenever the unobservable state is s. Thus this process will make a transition to a new state, and information about this new state becomes $\overline{\mu(y)_x}$ as Eq. (9), which is a probability distribution on the state space after moving forward by one unit. Let $\tilde{V}_n(\mu|y)$ be the expected value obtainable under the optimal policy, then

$$\tilde{V}_n(\mu|y) = \max_{0 \leq x \leq K} \left\{ -c(x) + \tilde{V}_{n-1}(\overline{\mu(y)_x}) \right\}. \tag{10}$$

Let $\tilde{V}_n(\mu)$ be an expected reward obtainable by employing the optimal policy, then the principle of optimality implies a recursive equation as

$$\tilde{V}_n(\mu) = \int_{-\infty}^{\infty} \tilde{V}_n(\mu|y) d\mu(y) \tag{11}$$

with $\tilde{V}_0(\mu) = \int_{-\infty}^{\infty} u(t) d\mu(t)$. In Eq. (10), $\mu(y)$ is posterior information about new state after improving information by using y from information process, and the expected value obtainable by the optimal policy is $\tilde{V}_{n-1}(\overline{\mu(y)_x})$.

By using an induction property with respect to n, the following properties are obtained under the assumptions treated in this section.

Proposition 2. *When μ satisfies a gradually condition, $\tilde{V}_n(\mu)$ is a nondecreasing function of μ, i.e. if $\mu \succeq \nu$, then $\tilde{V}_n(\mu) \geq \tilde{V}_n(\nu)$.*

If $\mu \succ \nu$, then $\tilde{V}_0(\mu) \geq \tilde{V}_0(\nu)$ by Lemma 8 since $u(t)$ is a non-decreasing non-negative function of t. When $\mu \succ \nu$, $\mu(y) \succ \nu(y)$ for any observation y by Lemma 10 and $\overline{\mu(y)_x} \succeq \overline{\nu(y)_x}$ for any amount x to expend by Lemma 16. These monotonic properties concerning posterior information imply that; if $\mu \succ \nu$, then $\overline{\mu(y)_x}(t) \succeq \overline{\nu(y)_x}(t)$ for any additional amount x of expenditure and any observed value y, and, therefore, Proposition 2 is obtained by employing an induction principle on n.

6. Conclusion

When we grasp the activity of the public sector as a management cycle of inputs, outputs and outcomes, we consider a rate of a population who satisfy a certain service as a barometer concerning the outcomes. Especially, a number of residents who satisfy a certain service is assumed to be distributed on $(-\infty, \infty)$ with distribution function $\Phi(x)$. In this paper, we considered an optimal expenditure problem as a sequential decision problem with Markovian transition. The problem is how much to spend to a public service to improve the outcomes within a range of the budget. For this expenditure problem on a partially observable Markov process, a gradually condition is introduced to a probability distribution since a state is not only reflected by transition but also changed by decisions. It is also possible to consider a monotonic property about the expected reward obtainable under the optimal policy by using this condition.

Acknowledgements The author wishes to thank the referees for their helpful comments. This research was partially supported by the Grant-in-Aid for Scientific Research of the Japan Society for the Promotion of Science and Technology, and the Grant-in-Aid for Research Project of Nomura Foundation for Academic Promotion.

References

1. F. De Vylder, Duality Theorem for Bounds in Integrals with Applications to Stop Loss Premiums, *Scandinavian Actuarial Journal*, 129–147, (1983).
2. Hedley, T. P. (1998), "Measuring Public Sector Effectiveness Using Private Sector Methods", *Public Productivity & Management Review*, **21** (3), 251–258.
3. M. Kijima and M. Ohnishi, Stochastic Orders and Their Applications in Financial Optimization, *Mathematical Methods of Operations Research*, **50**, 351–372, (1999).
4. T. Nakai, A Sequential Stochastic Assignment Problem in a Partially Observable Markov process, *Mathematics of Operations Research*, **11**, 230–240, (1986).
5. T. Nakai, An Optimal Selection Problem on a Partially Observable Markov process, In *Stochastic Modelling in Innovative Manufacturing*, Lecture Notes in Economics and Mathematical Systems **445**, (Eds. A. H. Christer, S. Osaki and L. C. Thomas), pp. 140–154, Springer-Verlag, Berlin, (1996).
6. T. Nakai, An Optimal Assignment Problem for Multiple Objects per Period – Case of a Partially Observable Markov process, *Bulletin of Informatics and Cybernetics*, **31**, 23–34, (1999).
7. T. Nakai, A Generalization of Multivariate Total Positivity of Order Two with an Application to Bayesian Learning Procedure, *Journal of Information & Optimization Sciences*, **23**, 163–176, (2002).
8. T. Nakai, Economy, Efficiency and Effectiveness, In *Policy Analysis in the Era of Globalization and Localization* (Eds. Research Project Group for Policy

Evaluation in Kyushu University), Kyushu University Press, 165–193, 2006.
9. T. Nakai, Properties of a Job Search Problem on a Partially Observable Markov Chain in a Dynamic Economy, *Computers & Mathematics with Applications*, The Special Issue on The Second Euro-Japanese Workshop on Stochastic Risk Modelling for Finance, Insurance, Production and Reliability, vol. 51, 189–198, 2006.
10. T. Nakai, A Sequential Expenditure Problem for Public Sector Based on the Outcome, *Recent Advances in Stochastic Operations Research* (Eds. T. Dohi, S. Osaki and K. Sawaki), World Scientific Publishing, 277–295, 2007.
11. S. M. Ross, *Stochastic Processes*, John-Wiley and Sons, New York, New York, 1983.

SEARCH FOR 90/150 CELLULAR AUTOMATA SEQUENCES WITH MAXIMUM MINIMUM-PHASE-SPACING

MASANORI FUSHIMI, TAKEHIRO FURUTA,* AND AKIHIRO ITO[†]

Graduate School of Mathematical Sciences and Information Engineering, Nanzan University, Seto-shi 489-0863, Japan
fushimi@nanzan-u.ac.jp

In this paper, we perform a phase shift analysis of the output sequences of maximum period-length 90/150 cellular automata(CA). We compute the minimum spacing among the phases of the output sequences of all the cells of an automaton. For the CA's with less than 25 cells, we perform an exhaustive search and find the best CA's, while for the CA's with more cells, we perform a random search to find a good CA's. The computational results are important for applications of the output sequences of CA's to built-in self-test of VLSI chips.

1. Introduction

Since the number of gates integrated into a VLSI chip are more than a million, it is practically impossible to test a chip for malfunction by exhaustively checking all the states that the chip can assume. So we check a chip by some mechanism of random sampling. A linear feedback shift register (LFSR) or a cellular automaton (CA) can be used to generate random test patterns. It is natural to include a testing mechanism into a chip in order to shorten the time required for test procedures. This method of testing is called "built-in self-test" (BIST). When we include a random pattern generator into a VLSI chip, it is desirable to make the area that the generator takes as compact as possible. In this respect, considerable interest has recently developed in the use of cellular automata.

The properties of CA's as pseudorandom sequence generators were originally studied by Wolfram [1], and a rule for naming CA's in one dimension with only nearest neighbor interactions was introduced by him as follows.

---

*Presently with Tokyo University of Science.
[†]Presently with Sony E.M.C.S., Inc.

	7	6	5	4	3	2	1	0
	111	110	101	100	011	010	001	000
Rule 90	0	1	0	1	1	0	1	0
Rule 150	1	0	0	1	0	1	1	0
	128	64	32	16	8	4	2	1

Each cell of a CA can assume the value either 0 or 1. The next state of a cell depends on itself and its nearest neighbors. We write the states of the cell and its neighbors at time t as 8 triplets in ascending (from right to left) binary order. The next state of the cell at time $t+1$ is written below the triplet in two lines for the two rules, rule 90 and rule 150, respectively. The group of 8 bits in each line is interpreted as a binary number between 0 and 255. This number is used as the name of the rule. In rule 90, the next state of a cell is the modulo-2 sum of the present states of the nearest neighbors. Rule 150 derives the next state of a cell as the modulo-2 sum of the present states of itself and both its neighbors. For the end cells, boundary conditions must be supplied. There are two possibilities, null or cyclic. In this paper, we assume null boundary conditions: the boundary always supplies a 0 from beyond the boundary into the modulo-2 sum that determines the state of the end cell.

It is known that CA's composed of cells with a single rule can rarely generate sequences of maximum possible period $2^n - 1$, where n is the number of cells, but hybrid CA's with certain combinations of rule 90 and rule 150 can generate maximum length cycles [2,3]. The properties of the sequences generated by maximum-length 90/150 (i.e. a hybrid of rules 90 and 150) CA's with null boundary conditions have extensively been studied [2,3,4,5]. For example, it is well-known that such CA's are isomorphic to LFSR's, i.e. the output sequence of any particular cell is the same as the output sequence of the corresponding LFSR except the phase shift. A most important property from the view point of applying the sequences to VLSI test is that every cell generates the same sequences with a phase shift, and phase shifts between adjacent cells are not constant, whereas these are constant (=1) in case of traditional LFSR sequences. Thus the output bit sequences from the various cells of a CA can be used in parallel as test stimuli. Before using these sequences in parallel, however, we must know the relative phase shifts between the output sequences from the various cells because the sequences with small relative phase shift can not be used in parallel. Methods of computing phase shifts have been studied by several researchers [2,4,5]. The objective of this paper is, using these methods and doing exhaustive search, to find CA's for which the minimum-phase-spacing, i.e. the minimum value of the relative phase shifts between various cells, is maximum or at least very large.

2. Designing the Maximum Period-Length 90/150 CA

Let the state of the k-th cell at time t of a n-cell 90/150 CA with null boundary conditions be denoted by $x_k(t)$. Then the state transition of this CA is described by the following equation:

$$x_k(t+1) = x_{k-1}(t) + c_k x_k(t) + x_{k+1}(t) \pmod{2},$$

where $c_k=0$ and $c_k=1$ correspond to rule 90 and rule 150, respectively. The boundary conditions are $x_0(t) = 0$ and $x_{n+1}(t) = 0$ for any t. In matrix notation, this can be written as a state transition equation

$$X(t+1) = AX(t)$$

over $\mathbb{F}_2$ (the Galois field with cardinality 2), where $X(t) = (x_1(t), x_2(t), ..., x_n(t))^{\mathrm{T}}$ is a column vector, and A is the tridiagonal matrix of the form

$$A = \begin{pmatrix} c_1 & 1 & & & & \\ 1 & c_2 & 1 & & & \\ & 1 & c_3 & 1 & & \\ & & \ddots & \ddots & \ddots & \\ & & & 1 & c_{n-1} & 1 \\ & & & & 1 & c_n \end{pmatrix}.$$

We call A the state transition matrix of the CA, and hereafter we write as $A = <c_1, c_2, \ldots, c_n>$ for simplicity.

It is well known that the sequence $X(t)$ attains the maximum possible period $2^n - 1$ if and only if the characteristic polynomial of A in $\mathbb{F}_2[x]$

$$p_n(x) = \det(xI + A) \tag{1}$$

is primitive. It is to be noted that $p_n(x)$ can be computed by the recursion

$$p_k(x) = (x + c_k)p_{k-1}(x) + p_{k-2}(x), \qquad p_0(x) = 1, \ p_{-1}(x) = 0, \tag{2}$$

which is obtained by expanding the determinant (1) using minors.

Bardell [2] proposed a method of designing a maximum-length 90/150 CA. His method chooses a state transition matrix A randomly, computes the characteristic polynomial $p_n(x)$, and check its primitivity. The probability that $p_n(x)$ is primitive is less than $2/n$ [3]. If it is not primitive, the process is repeated. Thus his method needs a trial and error and is inefficient when n is large. An efficient method, without a trial and error, for computing $c_1, c_2, \ldots, c_n$ given any primitive polynomial, say $p_n(x)$, in $\mathbb{F}_2[x]$ was given by Tezuka and Fushimi [3] based on a result by Mesirov and Sweet [6]. The outline of their method is as follows [3].

Step 1. Form the n by n matrix $B = (b_{ij})$ so that the following relationship with the given primitive polynomial $p_n(x)$ holds:

$$\sum_{j=1}^{n} b_{ij} x^{j-1} = x^{i-1} + x^{2i-1} + x^{2i} \pmod{p_n(x)}.$$

Step 2. Solve the following linear system over $\mathbb{F}_2$ to find $\boldsymbol{q} = (1, q_2, \ldots, q_n)^{\mathrm{T}}$:

$$B\boldsymbol{q} = (0, 0, \ldots, 0, 1)^{\mathrm{T}}.$$

Step 3. Let $p_{n-1}(x)$ be the polynomial of degree $n-1$ in $\mathbb{F}_2[x]$ obtained by taking only non-negative powers of the series

$$p_n(x)(x^{-1} + q_2 x^{-2} + \cdots + q_n x^{-n}).$$

Step 4. Using the Euclidean algorithm, compute $c_n, c_{n-1}, \ldots, c_1$ in this order so that the recurrence relation (2) holds.

The computational complexity of each step is $O(n^3)$, $O(n^3)$, $O(n^2)$, and $O(n^2)$, respectively.

3. Computing Phase Shifts of 90/150 CA Sequences

Let $j_1 = 0$, and j_k ($k = 2, \ldots, n$) be the relative phase shift of the output sequence of the k-th cell with respect to the output sequence of the 1st cell. Methods of computing these phase shifts have been proposed by several authors, e.g. [2,4,5]. We implemented the algorithms proposed by Sarkar [4] and by Cho et al. [5]. As for Sarkar's algorithm[4], its time complexity is $O(n^2 + n2^n)$ and space complexity is $O(n2^n)$.

Cho et al.'s algorithm[5] starts with an initial vector $X(0) = (1, 0, \ldots, 0)^{\mathrm{T}}$, and repeatedly multiply by the transition matrix A:

$$X(t) = AX(t-1), \quad t = 1, 2, \ldots$$

If $X(t)$ contains single 1, say the k-th element, then we put $j_k = -t$ (mod $2^n - 1$). This process is continued until we have computed all the j_k's. Thus the time complexity of this algorithm is $O(n2^n)$ and the space complexity is $O(n)$. As far as the space complexity is concerned, Cho et al.'s algorithm is much better than the Sarkar's algorithm. So we mainly use the Cho et al.'s in the following, but also use Sarkar's algorithm for relatively small n for comparison.

After computing all the phase shifts j_k ($k = 1, 2, ..., n$), which are in the interval $[0, 2^n - 2]$, we sort these to obtain the increasing sequence $0 = j(1) < j(2) < ... < j(n) < 2^n - 1$. Then we compute the minimum spacing $min\_spacing$ of this sequence as follows:

$$min\_spacing = \min\{j(k+1) - j(k) \mid k = 1, 2, \ldots, n\},$$

where $j(n+1) = 2^n - 1$.

We want to find, for each n, the CA whose $min\_spacing$ is maximum among all the n-cell 90/150 CA's. To do so, we must know all the primitive polynomials over $\mathbb{F}_2$ for degree n. A method of computing all the primitive polynomials from a primitive polynomial of the same degree is given in [7,8]. The number of the primitive polynomials of degree n over $\mathbb{F}_2$ is $\varphi(2^n - 1)/n$, where $\varphi(m)$ is the Euler's totient function.

4. Computational Results

Using the algorithms described in sections 2 and 3, we have computed the primitive polynomial with the maximum $min\_spacing$ for each degree n, $17 \leq n \leq 24$. The outline of the computational procedure for a given n is as follows:

Step 1. Compute all the primitive polynomials of degree n.
Step 2. Compute the transition matrix $A = < c_1, c_2, \ldots, c_n >$ for each primitive polynomial.
Step 3. Compute $min\_spacing$ for each transition matrix A, and find the one with the maximum $min\_spacing$.

Remark 4.1. We use the following two techniques to reduce the computational time in Step 3.

(1) It follows from the state transition equation, i.e. the first equation in Section 2, that $j_2 = -1$ if $c_1 = 0$, and that $j_{n-1} + 1 = j_n$ if $c_n = 0$, and the $min\_spacing$ is equal to 1 in both cases. Thus we need not compute $min\_spacing$ for any matrix A with c_1 or c_n equal to 0.
(2) Our objective is to find the primitive polynomial with the maximum $min\_spacing$ for each degree, and not to find the $min\_spacing$

of each primitive polynomial. Therefore we modify Cho et al.'s algorithm as follows. Rather than computing the *min_spacing* after all the phase shifts for a transition matrix A have been found, we compute the (temporal) *min_spacing* each time a new phase shift j_k for some k is found, using all the phase shifts found so far. If this (temporal) *min_spacing* happens to be smaller than the maximum of the *min_spacing*'s for all the primitive polynomials with the same degree examined up to that time, we skip the further computations for this transition matrix.

For the degrees $n \geq 25$, it is extremely time consuming to compute *min_spacing*'s for all the primitive polynomials (see Table 4). So we have chosen 1000 primitive polynomials randomly for each n and found the optimum polynomial among them. The computational results are shown in Table 1. Table 2 shows the diagonal elements $< c_1, c_2, \ldots, c_n >$ of the transition matrix A for the primitive polynomials shown in Table 1 for each n.

Tables 3, 4, and 5 show auxiliary information obtained through the computation. Table 3 shows the number and the percentage of primitive polynomials whose *min_spacing*'s are equal to 1 or 2. It shows that about 94% of primitive polynomials with degrees between 12 and 21 have the *min_spacing* 1 or 2. It means that randomly chosen primitive polynomials are not suitable for BIST, and a careful choice based on the computation of *min_spacing* such as done in this paper is essential. For example, Cho et al. [5] shows a computational result for a primitive polynomial with degree 32, but *min_spacing* for this polynomial is 1.

Tables 4 and 5 show the computational time for each degree[a]. We used a personal computer with Intel Pentium 4 CPU (3.0 GHz) and 1 GB memory for the computation. These tables show the exhaustive search for the higher degrees will become prohibitive.

5. Conclusion

Phase shift analysis is very important for applying the output sequences of cellular automata to built-in self-test of VLSI. We have computed phase shifts of maximum period-length 90/150 cellular automata sequences based on primitive polynomials over $\mathbb{F}_2$, and found primitive polynomials whose minimum spacings are maximum among all the primitive polynomials of

---

[a]The figures under Step 3 are based on the Cho et al.'s algorithm.

Table 1. The primitive polynomial (PP) with the maximum $min\_spacing$ (MMS) in each degree (deg.).

deg.	# PPs	PP with the maximum $min\_spacing$	MMS
17	7710	$1 + x + x^2 + x^3 + x^4 + x^5 + x^6 + x^{14} + x^{15} + x^{16} + x^{17}$	2787
18	7776	$1 + x^7 + x^8 + x^9 + x^{11} + x^{13} + x^{14} + x^{17} + x^{18}$	4657
19	27594	$1 + x^2 + x^3 + x^6 + x^7 + x^8 + x^{10} + x^{12} + x^{13} + x^{14} + x^{15} + x^{18} + x^{19}$	9205
20	24000	$1 + x + x^3 + x^4 + x^{10} + x^{11} + x^{12} + x^{13} + x^{14} + x^{18} + x^{20}$	20523
21	84672	$1 + x^2 + x^3 + x^4 + x^5 + x^7 + x^{11} + x^{13} + x^{14} + x^{15} + x^{21}$	33843
22	120032	$1 + x + x^2 + x^3 + x^5 + x^8 + x^{10} + x^{11} + x^{14} + x^{15} + x^{18} + x^{20} + x^{22}$	73913
23	356960	$1 + x^6 + x^7 + x^{10} + x^{11} + x^{14} + x^{15} + x^{16} + x^{17} + x^{18} + x^{19} + x^{22} + x^{23}$	152389
24	276480	$1 + x^3 + x^4 + x^5 + x^6 + x^9 + x^{14} + x^{16} + x^{18} + x^{20} + x^{24}$	224094
25	1296000	$1 + x^{10} + x^{12} + x^{15} + x^{17} + x^{21} + x^{25}$	288967*
26	1719900	$1 + x^2 + x^4 + x^5 + x^8 + x^{10} + x^{13} + x^{17} + x^{18} + x^{19} + x^{20} + x^{22} + x^{26}$	382874*
27	4202496	$1 + x^2 + x^4 + x^5 + x^7 + x^9 + x^{11} + x^{12} + x^{13} + x^{17}$ $+ x^{18} + x^{19} + x^{21} + x^{22} + x^{23} + x^{24} + x^{25} + x^{26} + x^{27}$	494737*
28	4741632	$1 + x^2 + x^3 + x^6 + x^{10} + x^{15} + x^{17} + x^{19} + x^{28}$	1548576*
29	18407808	$1 + x + x^2 + x^5 + x^8 + x^9 + x^{10} + x^{12} + x^{14} + x^{17} + x^{18} + x^{24} + x^{25} + x^{26} + x^{27} + x^{28} + x^{29}$	1748427*
30	17820000	$1 + x + x^5 + x^6 + x^7 + x^8 + x^{10} + x^{12} + x^{13} + x^{19} + x^{24} + x^{27} + x^{28} + x^{29} + x^{30}$	3453738*
31	69273666	$1 + x^2 + x^4 + x^{10} + x^{13} + x^{15} + x^{16} + x^{17} + x^{19} + x^{20} + x^{23} + x^{24} + x^{25} + x^{27} + x^{31}$	10697022*
32	67108863	$1 + x^3 + x^5 + x^9 + x^{13} + x^{18} + x^{19} + x^{27} + x^{28} + x^{29} + x^{30} + x^{31} + x^{32}$	7822043*

* : For the degrees 25 or larger, the optimum polynomial among randomly selected 1000 primitive polynomials and its $min\_spacing$ are shown.

Table 2. The diagonal elements of the state transition matrix A for the primitive polynomials shown in Table 1.

degree	$< c_1, \ldots, c_n >$
17	1000 1101 1001 0000 1
18	1010 1100 0101 0000 01
19	1010 0111 1001 1111 101
20	1000 1010 1100 0001 0101
21	1000 1011 1101 0110 1010 1
22	1011 1111 1011 1111 0100 01
23	1011 1110 1100 1110 1111 101
24	1010 1100 1101 1000 1111 0101
25	1010 0000 1111 1000 1101 0010 1
26	1000 1111 0111 0111 1001 1111 01
27	1010 1000 0000 0101 0000 0100 001
28	1010 1101 1100 1101 1111 0001 1101
29	1010 1101 1100 1101 0111 1111 0010 1
30	1010 0011 0111 0101 0101 1111 1000 01
31	1010 1000 1001 0000 1100 0101 1010 001
32	1000 1000 1010 0100 1100 1100 1011 0001

Table 3. The percentage of primitive polynomials whose $min\_spacing$'s are equal to 1 or 2 (We abbreviate $min\_spacing$ to MS in this table).

degree	# PPs	# MS=1	%	# MS=2	%	total (%)
12	144	110	76.39	23	15.97	92.36
13	630	475	75.40	119	18.89	94.29
14	756	570	75.40	139	18.39	93.78
15	1800	1359	75.50	327	18.17	93.67
16	2048	1551	75.73	363	17.72	93.46
17	7710	5775	74.90	1447	18.77	93.67
18	7776	5815	74.78	1463	18.81	93.60
19	27594	20709	75.05	5200	18.84	93.89
20	24000	17965	74.85	4483	18.68	93.53
21	84672	63534	75.04	15908	18.79	93.82

the same degree n, for $17 \leq n \leq 24$, and the best polynomials among randomly chosen 1000 primitive polynomials for the degrees more than 24. It is also shown that about 94% of the primitive polynomials have the minimum spacing 1 or 2, which are not suitable for application to BIST, and a careful choice of a primitive polynomial based on the phase shift analysis is essential.

For the purpose of applying the output sequences of CA's to BIST, it is also desirable to compute cross correlations between output sequences of different cells of a CA. The computational results will be published elsewhere.

Table 4. The computational time for the degrees between 17 and 24 (seconds).

degree	Step 1	Step 2	Step 3
17	0.24	0.42	24.51
18	0.29	0.45	55.04
19	1.16	1.97	368.48
20	1.24	2.02	562.37
21	4.23	6.03	4457.94
22	6.91	9.24	12534.92
23	20.35	30.49	72103.09
24	22.03	25.39	136965.05

Table 5. The computational time for the degrees between 25 and 32 (seconds): randomly selected 1000 primitive polynomials.

degree	Step 1	Step 2	Step 3
25	4.95	0.10	2008.81
26	5.10	0.11	3961.96
27	4.95	0.10	9745.89
28	4.91	0.09	14119.38
29	4.98	0.12	50751.73
30	4.92	0.15	97137.42
31	5.02	0.15	139713.70
32	5.00	0.18	370702.48

Acknowledgement The authors wish to thank the anonymous referee for a careful reading of the paper and useful suggestions. This research was partially supported by the Ministry of Education, Science, Sports and Culture, Grant-in-Aid for Scientific Research (A), Grant No. 17201037 (2005-2008).

References

1. Stephen Wolfram, Statistical mechanics of cellular automata, *Reviews of Modern Physics*, **55**, 601–644 (1983).
2. Paul H. Bardell, Analysis of cellular automata used as pseudorandom pattern generators, *Proceedings of the International Test Conference*, 762-768,(1990).
3. Shu Tezuka and Masanori Fushim, A method of Designing Cellular Automata as Pseudorandom Number Generators for Built-in Self-test for VLSI, *Contemporary Mathematicse*, **168**, 363-367,(1994).
4. Palash Sarkar, Computing Shifts in 90/150 Cellular Automata Sequences, *Finite Fields and Their Applications*, **9(2)**,175-186,(April,2003)
5. Sung-Jin Cho and Un-Sock Choi and Yoon-Hee Hwang and Han-Doo Kim and Yong-Soo Pyo and Kwang-Seok Kim and Seong-Hun Heo, Computing Phase Shifts of Maximum-Length 90/150 Cellular Automata Sequence, *Lecture Notes in Computer Science*, **3305**,31-39, Springer,(2004).
6. Jill P. Mesirov and Melvin M. Sweet, Continued Fraction Expansions of Rational Expressions with Irreducible Denominators in Characteristic 2, *Journal of Number Theory*, **27**,144-148,(1987).
7. Masanori Fushimi, *Random Numbers*, (University of Tokyo Press, 1989), (in Japanese)
8. Rudolf Lidl and Harald Niederreiter, *Introduction to Finite Fields and Their Applications*, (Cambridge University Press,1994), revised edition.

DIFFERENCE AND SIMILARITY BETWEEN MONANOVA AND OLS IN CONJOINT ANALYSIS

HIROMU KONO, HIROAKI ISHII* AND SHOGO SHIODE[†]

*Graduate School of Information Science and Technology, Osaka University,
2-1 Yamada-oka, Suita 565-0871, Japan,
hikouno@ist.osaka-u.ac.jp*

MONANOVA is a traditional method of conjoint analysis used for measuring the part worth value of factors in the total evaluation, exclusively using when evaluations is non-metrical data. The part worth values obtained by MONANOVA give an approximate comparison of each factor's contribution to the total evaluation, but it is impossible to utilize their contributions for statistical use since they are usually obtained by numerical solution. Moreover, they are not necessarily unique. In this paper, we first show the problems of MONANOVA and then propose a method to obtain its definite solution. With this, we also show the difference and similarity between MONANOVA and OLS which is typical method for measuring metric data.

1. Introduction

Conjoint analysis is a scaling method originally developed in mathematical psychology. Today it is used in many of the social sciences and applied sciences including marketing, product management, and operations research. It is used for measuring each factor's contribution to the whole evaluation of products consisting of some factors, for example, in testing customer acceptance of new product designs, in assessing the appeal of advertisements and in service design.

In conjoint analysis, many numbers of algorithms are used to estimate utility functions. Green and Srinivasan(1978) who developed conjoint analysis classified those estimation methods in three categories. First, they described methods that the total evaluation is assumed to be ordinal scaled. In that case estimation methods like MONANOVA(MONotone ANalysis Of Variance) or LINMAP can be used. Second, when it is assumed to be

---

*Graduate School of Information Science and Technology, Osaka University
[†]Faculty of Business Administration, Kobe Gakuin University

interval scaled, OLS(Ordinary Least Squares) regression techniques can be used. Third, for the paired comparison data in a choice context, the binary *Logit* or *Probit* model can be used.

With respect to MONANOVA, even with the method as the representative one of conjoint analysis, one cannot always measure theoretically the definite part worth values from the ranking data. Concerning this subject, Wittink and Cattin(1981) [12] showed that MONANOVA suffers from local optimum problems (which can be minimized if one uses multiple starting points). As a result their obtained scores cannot by default be applied in statistical methods. And they also showed very little difference between OLS and MONANOVA in terms of recovery of parameters, based on simulated rank order data. Thus, if OLS is about equally good as MONANOVA on ranks, there is no need to use MONANOVA. But since it is theoretically unacceptable to use OLS on ranks, it may for this reason be better to collect ratings, if interval-scales can be assumed. This particular undesirability of MONANOVA is well known; nevertheless, the method has been widely used, since it enables users to specify values of factors to a certain degree, simply from the ranking data. However, there are very few researchers to measure definite values directly by elaborating on MONANOVA.

Therefore, we would like to propose the method to obtain the formularization of MONANOVA's part worth values. This paper is organized as follows. In Section 2, we introduce MONANOVA and the traditional approach to obtain the part worth values. In Section 3, we introduce OLS. In Section 4, we propose the method to obtain the formularization of MONANOVA's part worth values and illustrate them with an example. Finally in Section 5, we summarize our results.

2. MONANOVA

In this section, we introduce MONANOVA. We use the following notation throughout the paper. Y is the order preserving transformation of ordinal scale, where m is the number of samples, and T means transpose,

$$Y = [Y_1, Y_2, \cdots, Y_m]^T \tag{1}$$

D is the 0-1 design matrix indicating each level of factors of samples, where n is the number of levels for all factors

$$D = \begin{bmatrix} D_1 \\ D_2 \\ \vdots \\ D_j \\ \vdots \\ D_m \end{bmatrix} = \begin{bmatrix} d_{11} & d_{21} & \cdots & d_{i1} & \cdots & d_{n1} \\ d_{12} & d_{22} & \cdots & d_{i2} & \cdots & d_{n2} \\ \vdots & \vdots & \ddots & \vdots & \ddots & \vdots \\ d_{1j} & d_{2j} & \cdots & d_{ij} & \cdots & d_{nj} \\ \vdots & \vdots & \ddots & \vdots & \ddots & \vdots \\ d_{1m} & d_{2m} & \cdots & d_{im} & \cdots & d_{nm} \end{bmatrix} \quad (2)$$

It follows that $\sum_{i=1}^{n} d_{ij} = l$, respectively, at $j = 1 \cdots m$, where l is the number of factors. B is the set of part worth values to be estimated and $B_j (j = 1 \cdots n)$ express the importance of the factor j.

$$B = [B_1, B_2, \cdots, B_n]^T \quad (3)$$

$X(B)$ is the equation of *the conjoint model*.

$$X(B) = [X_1(B), X_2(B), \cdots, X_m(B)]^T \quad (4)$$

We obtain the part worth values B by minimizing the goodness of fit criterion S under the restriction of *conjoint model* $X(B)$ and Y. In MONANOVA, we use Kruskal's Stress S (5) for the goodness of fit criterion and additive model $X_i(B)$ (6) for *conjoint model*.

$$S(B) = \min_{B} \left[\sqrt{\frac{\sum_{i=1}^{m}(Y_i - X_i(B))^2}{\sum_{i=1}^{m}(X_i(B) - \bar{X})^2}} \right], \quad (\bar{X} = \frac{1}{m}\sum_{i=1}^{m} X_i(B)) \quad (5)$$

$$X_i(B) = D_i B \quad (6)$$

The traditional approach to obtain the part worth value B is the following iterative algorithm.

Algorithm

Step 1: Choose the initial value of $B^{(k)}$ with random and set $k = 0$. Select a convergence parameter $\epsilon > 0$.
Step 2: Calculate the next equation (7) and get $B^{(k+1)}$.

$$B_j^{(k+1)} = B_j^{(k)} - \alpha \frac{\partial S(B^{(K)})}{\partial B_j} \quad (7)$$

$$S(B) = \sqrt{\frac{E(B)}{U(B)}} \quad (E(B) = \sum_{i=1}^{m}(Y_i - X_i)^2, U(B) = \sum_{i=1}^{m}(X_i - \bar{X})^2) \quad (8)$$

The partial derivative of S with respect to B_j is

$$\frac{\partial S(B)}{\partial B_j} = \frac{1}{2E}(\sqrt{\frac{U}{E}} \cdot \frac{\partial E}{\partial B_j} - \sqrt{\frac{E}{U}} \cdot \frac{\partial U}{\partial B_j}) = \frac{1}{2ES}(\frac{\partial U}{\partial B_j} - S^2 \frac{\partial E}{\partial B_j}) \quad (9)$$

$$\begin{cases} \dfrac{\partial E(B)}{\partial B_j} = 2\sum_{i=1}^{m} d_{ij}(Y_i - D_i B) \\ \dfrac{\partial U(B)}{\partial B_j} = 2\sum_{i=1}^{m}(D_i B - \bar{D}B) \end{cases}$$

Step 3: If $S(B^{(K)}) - S(B^{(K+1)}) < \epsilon$, then stop the iteration process as $B^{(k)}$ is a minimum point of $S(B)$. Otherwise, set $k = k+1$ and go to Step 2.

However using this numerical approach, their solution is not correct and unique and depended on the starting point $B^{(0)}$. Moreover, they suffer from local optimum problems. As a result we cannot apply especially them as statistical methods.

3. OLS

When it is assumed that the dependent variable is interval scaled, OLS(Ordinary Least Squares) regression techniques can be used. In OLS, we use the following equation (10) for the goodness of fit criterion.

$$S = \min_{B}[\sum_{i=1}^{m}\{X_i(B) - Y_i\}^2] \quad (10)$$

When we use the additive conjoint model $X(B) = DB$, B satisfies (13).

$$S = \min_{B}[\sum_{i=1}^{m}(D_i B - Y_i)^2] \quad (11)$$

$$\frac{\partial S}{\partial B} = 2D^T(DB - Y) = 0 \quad (12)$$

$$D^T DB = D^T Y \quad (13)$$

The traditional textbook approach to avoiding the dummy trap problem is to delete a category from each qualitative variable. We transform the matrix D to the matrix D^* using a full rank one-dimensional quantification method similar to Hayashi's quantification method-I [3], where D^* is the matrix obtained by removing the one category's column of each factor except one from D.

$$B' = (D^{*T}D^*)^{-1}D^{*T}Y \qquad (14)$$

And we obtain the part worth values $\hat{B}$ for each factor were centered around zero from B'.

$$B' = \begin{bmatrix} B_1 \\ \vdots \\ B_n \end{bmatrix} \rightarrow \hat{B}' = \begin{bmatrix} \bar{B} \\ b_1 \\ \vdots \\ b_n \end{bmatrix} \qquad (15)$$

Here, $\bar{B}$ is the average of levels for all factors. $\bar{B} = \dfrac{1}{m}\sum_{i=1}^{n} B_i$.
$b_i = B_i - u_k$ (k is the number of factor which level i belongs to. u_k is the average of levels B_i which belong to factor k)

Since the traditional approach cannot get the equation of $\hat{B}$, we propose the method with using the dummy variable coefficients. Where l is the number of factors. We define $(n \times l)$ 0-1 design matrix $C = (c_{ij})$ to indicate the dummy variable coefficients, as such

$$C = \begin{bmatrix} C_1 \\ \vdots \\ C_l \end{bmatrix} = \begin{bmatrix} c_{11} & c_{21} & \cdots & c_{n1} & \cdots & c_{n1} \\ \vdots & \vdots & \ddots & \vdots & & \vdots \\ c_{1k} & c_{2k} & \cdots & c_{ik} & \cdots & c_{nk} \\ \vdots & \vdots & & \vdots & \ddots & \vdots \\ c_{1l} & c_{2l} & \cdots & c_{n1} & \cdots & c_{nl} \end{bmatrix} \qquad (16)$$

If a level i belongs to the factor k, $c_{ik} = 1$. If not, $c_{ik} = 0$. We denote the I is the $(m \times 1)$ vector with all elements being 1, $I = [1, 1, \cdots, 1]^T$, O is the $(l \times 1)$ vector with all elements being $O = [0, \cdots, 0]^T$ and $b = [b_1, \cdots, b_i, \cdots, b_n]$. We use the

$$\hat{D} = \begin{bmatrix} I & D \\ O & C \end{bmatrix}, \hat{B} = \begin{bmatrix} \bar{B} \\ b \end{bmatrix}, \hat{Y} = \begin{bmatrix} Y \\ O \end{bmatrix} \qquad (17)$$

$$\hat{D}^T \hat{D} \hat{B} = \hat{D}^T \hat{Y} \qquad (18)$$

We can obtain $\hat{B}$, as follows.

$$\hat{B} = (\hat{D}^T\hat{D})^{-1}\hat{D}^T\hat{Y} \tag{19}$$

Both (14) and (19) can be translated into the following equations.

$$\begin{cases} m\bar{B} + \sum_{i=1}^{n}\sum_{j=1}^{m} d_{ij}b_i = \sum_{j=1}^{m} Y_j \\ \sum_{j=1}^{m} d_{kj}\bar{B} + \sum_{i=1}^{n} d_{kj}d_{ij}b_i = \sum_{j=1}^{m} d_{kj}Y_j \end{cases} \tag{20}$$

So the $\hat{B}$ of (19) is the same as $\hat{B}'$ of (14). Using (19), we can obtain the OLS solution easily.

4. Proposal Methods

4.1. *Proposal method*

Using the following equations, we obtain the part worth values $\hat{B}$ without using iterative algorithm.

$$\hat{B} = \frac{1}{1-\lambda}(\hat{D}^T\hat{D})^{-1}\hat{D}^T(\hat{Y} - \lambda\bar{Y}\hat{I}) \tag{21}$$

$$\lambda = \frac{\hat{Y}^T\{\hat{Y} - \hat{D}(\hat{D}^T\hat{D})^{-1}\hat{D}^T\hat{Y}\}}{\hat{Y}^T(\hat{Y} - \bar{Y}I)} \tag{22}$$

$$\hat{D} = \begin{bmatrix} I & D \\ O & C \end{bmatrix}, \hat{B} = \begin{bmatrix} \bar{B} \\ b \end{bmatrix}, \hat{Y} = \begin{bmatrix} Y \\ O \end{bmatrix}, \hat{I} = \begin{bmatrix} I \\ O \end{bmatrix} \tag{23}$$

We define the equation of $g(\lambda, B)$ (24) and $G(\lambda)$ (25), as follows

$$g(\lambda, B) = \sum_{i=1}^{m}\{(X_i - Y_i)^2 - \lambda(X_i - \bar{X})^2\} \tag{24}$$

$$G(\lambda) = \min_B[g(\lambda, B)] = g(\lambda, B^*) \tag{25}$$

Theorem 4.1. *If*

$$G(\lambda^*) = 0 \tag{26}$$

then

$$\lambda^* = \min_B[S(B)] = S(B^*) \tag{27}$$

Proof. For $\lambda_1 \leq \lambda_2$, we have

$$G\{t\lambda_1 + (1-t)\lambda_2\}$$
$$= \min_B \sum_{i=1}^m [(X_i - Y_i)^2 - \{t\lambda_1 + (1-t)\lambda_2\}(X_i - \bar{Y})^2]$$
$$= \min_B \sum_{i=1}^m [t(X_i - Y_i)^2 + (1-t)(X_i - Y_i)^2 - \{t\lambda_1 + (1-t)\lambda_2\}(X_i - \bar{Y})^2]$$
$$= \min_B \sum_{i=1}^m t\{(X_i - Y_i)^2 - \lambda_1(X_i - \bar{Y})^2\}$$
$$+ \min_B \sum_{i=1}^m (1-t)\{(X_i - Y_i)^2 - \lambda_2(X_i - \bar{Y})^2\} \leq tG(\lambda_1) + (1-t)G(\lambda_2). \quad (28)$$

and

$$G(\lambda_1) = \min_B \sum_{i=1}^m [(X_i - Y_i)^2 - \lambda_1(X_i - \bar{Y})^2] = \sum_{i=1}^m [(X_i^* - Y_i)^2 - \lambda_1(X_i^* - \bar{Y})^2]$$
$$\geq \sum_{i=1}^m [(X_i^* - Y_i)^2 - \lambda_2(X_i^* - \bar{Y})^2] \geq \min_B \sum_{i=1}^m [(X_i - Y_i)^2 - \lambda_2(X_i - \bar{Y})^2]$$
$$= G(\lambda_2) \qquad (29)$$

Hence, $G(\lambda)$ is monotone decreasing and concave function of λ. Furthermore,

$$G(\lambda = 0) = \min_B \sum_{i=1}^m (X_i - Y_i)^2 \geq 0 \qquad (30)$$

$$G(\lambda = -\infty) = \lim_{\lambda \to -\infty} \left[\min_B \sum_{i=1}^m \{(1-\lambda)(X_i - \frac{Y_i - \lambda\bar{Y}}{1-\lambda})^2 - \lambda(Y_i - \bar{Y})^2\} \right]$$
$$< 0 \qquad (31)$$

so there exists a unique λ such that $G(\lambda) = 0$; that is,

$$G(\lambda^*) = g(\lambda^*, B^*) = \sum_{i=1}^m [(X_i(B^*) - Y_i)^2 - \lambda^*(X_i(B^*) - \bar{Y})^2] \qquad (32)$$

Since

$$0 = \sum_{i=1}^m [(X_i(B^*) - Y_i)^2 - \lambda^*(X_i(B^*) - \bar{Y})^2]$$
$$\leq \sum_{i=1}^m [(X_i(B) - Y_i)^2 - \lambda^*(X_i(B) - \bar{Y})^2] \qquad (33)$$

for any B, and

$$\lambda^* = \frac{\sum_{i=1}^{m}(X_i(B^*) - Y_i)^2}{\sum_{i=1}^{m}(X_i(B^*) - \bar{Y})^2} \leq \frac{\sum_{i=1}^{m}(X_i(B) - Y_i)^2}{\sum_{i=1}^{m}(X_i(B) - \bar{Y})^2} = S(B) \qquad (34)$$

We have that B^* is an optimal solution of $S(B)$.

Theorem 4.2. *We denote the average of X with $\bar{X}$ and that of Y with $\bar{Y}$. When we wish to minimize $S(B)$ with respect to B, we have*

$$\bar{X} = \bar{Y}. \qquad (35)$$

Proof.

$$X_i = \bar{X} + x_i, \quad Y_i = \bar{Y} + y_i \quad (\bar{X} = \frac{1}{m}\sum_{i=1}^{m}X_i, \quad \bar{Y} = \frac{1}{m}\sum_{i=1}^{m}Y_i) \qquad (36)$$

Using (36), we calculate (24).

$$g(\lambda, B) = \sum_{i=1}^{m}\{(X_i - Y_i)^2 - \lambda(X_i - \bar{X})^2\}$$

$$= \sum_{i=1}^{m}\{(\bar{X} - \bar{Y})^2 + 2(x_i - y_i)(\bar{X} - \bar{Y}) - (x_i - y_i)^2 - \lambda x_i^2\}$$

$$= 3m(\bar{X} - \bar{Y})^2 + \sum_{i=1}^{m}\{(x_i - y_i)^2 - \lambda x_i^2\} \geq \sum_{i=1}^{m}\{(x_i - y_i)^2 - \lambda x_i^2\}$$

$$= \sum_{i=1}^{m}\{(X_i - Y_i)^2 - \lambda(X_i - \bar{Y})^2\} \qquad (37)$$

From Theorem 4.1, when we minimize $S(B)$ with respect to B, $G(\lambda) = 0$. So $\bar{Y} = \bar{X}$.

$$G(\lambda) = \min_{B} g(\lambda, B)$$

$$\geq \min_{B} \sum_{i=1}^{m}\{(X_i(B) - Y_i)^2 - \lambda(X_i(B) - \bar{Y})^2\} = 0 \qquad (38)$$

Theorem 4.3. *If $X^2(B)$ is a convex function of b_i, the minimizer b_i of $g(\lambda, B)$ is the solution of*

$$\sum_{i=1}^{m}\frac{\partial X_i(B)}{\partial b_j}\{(X_i(B) - Y_i) - \lambda(X_i(B) - \bar{Y})\} = 0 \qquad (39)$$

Proof.

$$\min_B g(\lambda, B)$$
$$\geq \min_B \sum_{i=1}^{m} \{(X_i(B) - Y_i)^2 - \lambda(X_i(B) - \bar{Y})^2\}$$
$$= (1-\lambda)\min_B\{\sum_{i=1}^{m}(X_i(B) - \frac{Y_i - \lambda\bar{Y}}{1-\lambda})^2\} - \lambda\sum_{i=1}^{m}(Y_i - \bar{Y})^2 \quad (40)$$

$$tg(\lambda, B_{(b_i=a_1)}) + (1-t)g(\lambda, B_{(b_i=a_2)}) \geq g(\lambda, B_{(b_i=ta_1+(1-t)a_2)}) \quad (41)$$

$g(\lambda, B)$ is a convex function of b_i, so, when we minimize $g(\lambda, B)$ with respect to b_i, $\dfrac{\partial S(\lambda, B)}{\partial b_i} = 0$.

From Theorems 4.1–4.3, the minimizer B of $S(B)$ is the solutions of

$$(DB - Y)^T(DB - Y) - \lambda(DB - \bar{Y}I)^T(DB - \bar{Y}I) = 0 \quad (42)$$
$$D^T\{(DB - Y) - \lambda(DB - \bar{Y}I)\}] = 0 \quad (43)$$

that is, the solution of

$$Y^T(Y - DB) = 0 \quad (44)$$
$$D^TDB = \frac{1}{1-\lambda}D^T(Y - \lambda\bar{Y}I) \quad (45)$$

When $Y' = \frac{1}{1-\lambda}(Y - \lambda\bar{Y}I)$, (45) is $D^TDB = D^TY'$. So we obtain B by the method similar to (19).

$$\hat{D} = \begin{bmatrix} I & D \\ O & C \end{bmatrix}, \hat{B} = \begin{bmatrix} \bar{B} \\ b \end{bmatrix}, \hat{Y} = \begin{bmatrix} Y \\ O \end{bmatrix}, \hat{I} = \begin{bmatrix} I \\ O \end{bmatrix} \quad (46)$$

$$\hat{Y}^T(\hat{Y} - \hat{D}\hat{B}) = 0 \quad (47)$$
$$\hat{D}^T\hat{D}\hat{B} = \frac{1}{1-\lambda}\hat{D}^T(\hat{Y} - \lambda\bar{Y}\hat{I}) \quad (48)$$

We obtain the part worth values $\hat{B}$ as the following equations.

$$\hat{B} = \frac{1}{1-\lambda}(\hat{D}^T\hat{D})^{-1}\hat{D}^T(\hat{Y} - \lambda\bar{Y}\hat{I}) \quad (49)$$

$$\lambda = \frac{\hat{Y}^T\{\hat{Y} - \hat{D}(\hat{D}^T\hat{D})^{-1}\hat{D}^T\hat{Y}\}}{\hat{Y}^T(\hat{Y} - \bar{Y}I)} \quad (50)$$

Setting $\alpha = 1/(1-\lambda)$, we obtain

$$\alpha = \frac{1}{1-\lambda} = \frac{(Y-\bar{Y}\hat{I})^T(Y-\bar{Y}\hat{I})}{\{\hat{D}(\hat{D}^T\hat{D})^{-1}\hat{D}^T\hat{Y}-\bar{Y}\hat{I}\}^T\{\hat{D}(\hat{D}^T\hat{D})^{-1}\hat{D}^T\hat{Y}-\bar{Y}\hat{I}\}} \quad (51)$$

$$\begin{bmatrix} \bar{B}+(\alpha-1)\bar{Y} \\ b \end{bmatrix} = \alpha(\hat{D}^T\hat{D})^{-1}\hat{D}^T\hat{Y} \quad (52)$$

Using the part worth values $\hat{B}_{OLS} = (\hat{D}^T\hat{D})^{-1}\hat{D}^T\hat{Y}$ by OLS, the part worth value $\hat{B}_{MONA}$ by MONANOVA is derived to the following simple equation.

$$\begin{bmatrix} \bar{B}+(\alpha-1)\bar{Y} \\ \hat{b}_{MONA} \end{bmatrix} = \alpha\hat{B}_{OLS} \quad (53)$$

$$\alpha = \frac{\sum_{i=1}^{m}(Y_i-\bar{Y})^2}{\sum_{i=1}^{m}(D_i\hat{B}_{OLS}-\bar{Y})^2} \quad (54)$$

4.2. Numerical example

In this subsection, we show a numerical example. In Table 1, five different kinds of car from 1 to 5 are given. These cars consist of two factors (Color and Style). Y is scored higher according to a consumer's preference ranking. We want to obtain the commercial values of each piece of car by analyzing the two factors from the consumers' preference ranking. By adopting MONANOVA, these commercial values correspond to the part worth value B.

Table 1. Consumer's preference example of five cars.

No.	Color			Style		Score Y
	pink	white	blue	Sedan	Sports	
1	1	0	0	1	0	1
2	0	1	0	0	1	6
3	0	0	1	1	0	3
4	1	0	0	0	1	2
5	0	1	0	1	0	4
	B_1	B_2	B_3	B_4	B_5	

We try to compute the part worth value B of this sample using the proposed method in the previous subsection. From Table 1, D, B, Y, C is

$$D = \begin{bmatrix} 1 & 0 & 0 & 1 & 0 \\ 0 & 1 & 0 & 0 & 1 \\ 0 & 0 & 1 & 1 & 0 \\ 1 & 0 & 0 & 0 & 1 \\ 0 & 1 & 0 & 1 & 0 \end{bmatrix}, C = \begin{bmatrix} 1 & 1 & 1 & 0 & 0 \\ 0 & 0 & 0 & 1 & 1 \end{bmatrix}, B = \begin{bmatrix} B_1 \\ B_2 \\ B_3 \\ B_4 \\ B_5 \end{bmatrix}, Y = \begin{bmatrix} 1 \\ 6 \\ 3 \\ 2 \\ 4 \end{bmatrix}. \quad (55)$$

By using equations (46), we change (55) to $\hat{D}, \hat{B}, \hat{Y}, \hat{I}$.

$$\hat{D} = \begin{bmatrix} 1 & 1 & 0 & 0 & 1 & 0 \\ 1 & 0 & 1 & 0 & 0 & 1 \\ 1 & 0 & 0 & 1 & 1 & 0 \\ 1 & 1 & 0 & 0 & 0 & 1 \\ 1 & 0 & 1 & 0 & 1 & 0 \\ 0 & 1 & 1 & 1 & 0 & 0 \\ 0 & 0 & 0 & 0 & 1 & 1 \end{bmatrix}, \hat{B} = \begin{bmatrix} \bar{B} \\ b_1 \\ b_2 \\ b_3 \\ b_4 \\ b_5 \end{bmatrix}, \hat{Y} = \begin{bmatrix} 1 \\ 6 \\ 3 \\ 2 \\ 4 \\ 0 \\ 0 \end{bmatrix}, \hat{I} = \begin{bmatrix} 1 \\ 1 \\ 1 \\ 1 \\ 1 \\ 0 \\ 0 \end{bmatrix} \quad (56)$$

The $\bar{Y}$, average of Y, is $\bar{Y} = 3.2$. Now, we get the $\hat{B}_{OLS}$ by using (19).

$$\hat{B}_{OLS} = (\hat{D}^T \hat{D})^{-1} \hat{D}^T \hat{Y} = \begin{bmatrix} 3.41667 \\ -1.91667 \\ 1.58333 \\ 0.33333 \\ -0.75 \\ 0.75 \end{bmatrix} \quad (57)$$

By using the equations (50), (57) and (55), we can calculate α,

$$\alpha = \frac{\sum_{i=1}^{m}(Y_i - \bar{Y})^2}{\sum_{i=1}^{m}(D_i \hat{B}_{OLS} - \bar{Y})^2} = 1.01708 \quad (58)$$

and $\hat{B}$ is given by (49) and (55), such that

$$\begin{bmatrix} \bar{B} + (\alpha - 1)\bar{Y} \\ b \end{bmatrix} = \alpha \hat{B}_{OLS} = 1.01708 \begin{bmatrix} 3.41667 \\ -1.91667 \\ 1.58333 \\ 0.33333 \\ -0.75 \\ 0.75 \end{bmatrix} \tag{59}$$

$$\rightarrow \hat{B} = \begin{bmatrix} \bar{B} \\ b \end{bmatrix} = \begin{bmatrix} 3.4204 \\ -1.9496 \\ 1.6105 \\ 0.3390 \\ -0.7629 \\ 0.7629 \end{bmatrix} \tag{60}$$

By the way, we computed the B_{algo} using conventional algorism (7) (start from $\bar{B} = [1, 0, 0, 0, 0]^T$), and calculate $\hat{B}_{algo}$ that is the part worth value for each factor were centered around zero.

$$B_{algo} = \begin{bmatrix} -0.381 \\ 3.179 \\ 1.907 \\ 1.089 \\ 2.615 \end{bmatrix} \rightarrow \hat{B}_{algo} = \begin{bmatrix} 3.4204 \\ -1.9496 \\ 1.6105 \\ 0.3390 \\ -0.7629 \\ 0.7629 \end{bmatrix} \tag{61}$$

$\hat{B} = \hat{B}_{algo}$. Thus, we see the part worth value of MONANOVA can be obtained by using the proposed method.

Table 2. The part worth value of sample.

	Pink	White	Blue	Sedan	Sports	
MONANOVA	-1.9496	1.6105	0.3390	-0.7629	0.7629	3.4204
OLS	-1.91667	1.58333	0.33333	-0.75	0.75	3.41667
	B_1	B_2	B_3	B_4	B_5	$\bar{B}$

5. Conclusion

In this paper, we have proposed the method to obtain the formula of MONANOVA's part worth value. With this formula, we can easily obtain the part worth value and prevent the local optimum problems. We also have shown the difference and similarity between OLS and MONANOVA in terms of recovery of parameters. Consequently, the score of MONANOVA is the spread score of OLS with the ratio of the total variance of Y to that of X. When we use their score for measuring each factor's contribution to the whole evaluation, they are showed to be very little difference, and OLS is about equally good as MONANOVA on ranks. Therefore, there is no need to use MONANOVA.

As future studies, we would like to show the difference and similarity of MONANOVA and the other method of conjoint analysis. The score of LINMAP is known similar to that of MONANOVA.[11] We show similarity between LINMAP and MONANOVA more explicitly, though it is suggested by Green.

References

1. J. Green, P. Carroll and S. Goldberg (1981). A general approach to product design optimization via conjoint analysis. *Journal of Marketing*, Vol. 43, pp. 17–25.
2. P. Green and V. Srinivasan (1978). Conjoint analysis in consumer research: issues and outlook. *Journal of Consumer Research*, Vol. 5, pp. 103–123.
3. C. Hayasi (1968). One ddimensional quantification and multidimensional quantification. *Annals of the Japan Association for Philosophy of Science*, Vol. 3, pp. 115–120.
4. H. Noguchi and H. Ishii (2000). Methods for determining the statistical part worth value of factors in conjoint analysis. *Mathematical and Computer Modelling*, Vol. 31, pp. 261–272.
5. J. Iwase T. Hasegawa I. Ibaraki, H. Ishii and H. Mine (1976). Algorithms for quadratic fractional programming problems. *Journal of the Operations Research Society of Japan*, Vol. 19, No. 2, pp. 228–244.
6. J.B. Kruskal (1964). Multidimensional scaling by optimizing goodness of fit to a nonmetric hypothesis. *Psychometrika*, Vol. 29, No. 1, pp. 1–7.
7. J.B. Kruskal (1965). Analysis of factorial experiments by estimating monotone transformations of the data. *J. Royal Statist.*, Series B, Vol. 27, No. 2, pp. 251–263.
8. J.B. Kruskal and F.J. Carmone (1969). Monanova, a fortran program for monotone analysis of variance(non-metric analysis of factorial experiments). *Behav. Sci.*, Vol. 14, pp. 165–166.
9. Bryan Orme. Sawtooth Software technical papers: General Conjoint Analysis. Sawtooth Software, 2005. http://www.sawtoothsoftware.com/techpap.shtml.
10. P.A. Green and V.R. Rao (1971). Conjoint measurement for quantifying judgemental data. *Journal of Marketing Research*, Vol. 8, pp. 355–363.

11. W. Dinkelbach (1967). On nonlinear fractional programming. *Management Science*, Vol. 13, No. 7, pp. 492–498.
12. D.R. Wittink and P. Cattin (1981), Alternative estimation methods for conjoint analysis: a Monte Carlo study, *Journal of Marketing Research*, Vol. 18, pp. 101–106.

Part B Stochastic Modeling

A DATUM SEARCH GAME AND AN EXPERIMENTAL VERIFICATION FOR ITS THEORETICAL EQUILIBRIUM

RYUSUKE HOHZAKI[†] AND YOSHIHIKO IDA[‡]

[†]*Department of Computer Science, National Defense Academy,*
1-10-20 Hashirimizu, Yokosuka, 239-8686 Japan,
hozakicc.nda.ac.jp
[‡]*Air Staff Office, Ministry of Defense,*
5-1 Ichigaya-honmura, Sinjuku-ku, Tokyo, 162-8804 Japan

A so-called datum search game is a kind of two-person zero-sum search games. Two players, a searcher and a target, start playing the game by the information of a point of the target. For the game, Washburn and Hohzaki already proposed a theoretical method to derive an approximation of an equilibrium solution, which is represented by a motion for the target and a plan of distributing search resources for the searcher in a continuous search space. Recently, there have been trials that verify theoretical solutions of games by some experiments involving human factors, especially in Experimental Economics. We make a simulator consisting of two PCs, on which two operators play roles of the target and the searcher through the simulation of the datum search game. After analyzing data on actions of two operators, we obtain the experimental results, which are consistent with a theoretical equilibrium.

1. Introduction

Since von Neumann and Morgenstern published a book entitled "Theory of Games and Economic Behavior" [5], game theory has been applied to a variety of practical fields, especially, to micro-economics. Game theory becomes popular now owning to Nobel prize winners such as Harsanyi, Nash, Selten and others. However we often observe that theoretical solutions or theoretical strategies of the games do not necessarily correspond to decision makings of human in economic field or other real world. Recently, many researchers try to inspect the theoretical solutions or optimal strategies of the game from an experimental point of view, that is, through experiments with human factors, as we see in experimental economics.

The above situation is fact in search theory. Search theory involves many practical actions of human like search-and-rescue operations in the ocean and therefore theoretical results for search games need to be reviewed from a new view point through simulations or experiments with human factors. This paper deals with a so-called datum search game, which is a kind of two-person zero-sum search games. We examine theoretical results obtained in past researches by some simulations. In the search-and-rescue activities or military operations, where a searcher wants to detect a target, we call the exposed target position "datum point", and information including the datum point, time or others "datum information". Being motivated by datum information, a searcher and a target start a datum search game.

The datum search game has many applications such as the search-and-rescue activity and the military operation in the ocean. Koopman [9] gets together the results of the naval operations research by the U.S. Navy in the Second World War. He studies a so-called datum search but not a game, where a target takes a diffusive motion after randomly selecting his course from a datum point. Direct applications of the datum search model could be military operations such as anti-submarine warfare (ASW). Danskin [2], Baston and Bostock [1] and Garnaev [4] studies the ASW search games. Other than the datum search game, generalized models of search games have been discussed by Meinardi [10], Washburn [13], Nakai [11], Thomas and Washburn [12], Eagle and Washburn [3] and Kikuta [8] as single-stage or multi-stage search games.

The search games with a moving target are approximately classified into two types of models. One is called search-and-evasion game [13], where both of a target and a searcher take moving strategies, and the other is called search allocation game (SAG) [6], where a target moves but a searcher distributes a limited amount of searching resources to detect the target. There are not so many researches for the SAG. Because of the difficulty of solving the game, they often simplify their models to solve the problems, e.g. by the neglect of the continuity of the target motion in a search space [2]. On the other hand, Washburn or Hohzaki [14,7] discuss the datum search game taking account of practical constraints about energy consumption and continuous motion for the target. However they could not exactly solve the game but propose an upper bound and a lower bound on the value of the game.

In this paper, we review theoretical results of Washburn and Hohzaki and show their solution in Section 2. We explain functions of our simulator consisting we made, with which two operators play roles as a searcher and a target, in Section 3. In Section 4, we show our results of simulations in terms of the characteristics of the operators' decision makings and compare them with Washburn and Hohzaki's theoretical results.

2. Datum Search Game and Theoretical Results

2.1. *A model of datum search game and formulation*

Here, according to Washburn and Hohzaki's studies [14,7], we review a mathematical model of a datum search game in a two-dimensional space and their theoretical results. We consider the following SAG, in which a searcher and a target participate, in a two-dimensional continuous space.

(1) A game is played in a two-dimensional continuous Euclidean space R^2. A datum point is placed at the origin at time $t = 0$.
(2) After the datum point is reported to a searcher, he is dispatched to the datum area. He starts the datum search after time lag τ and continues it until time T. The searcher can distribute the total amount ρ of divisible searching resources per unit time in the search space.
(3) The target is just at the origin at time $t = 0$ and moves on a plane R^2. The usage of speed v spends energy $\mu(v)$ per unit time. The function $\mu(v)$ is assumed to be convex and monotone increasing for v. The speed over maximum speed V_c is prohibited. The target possesses energy e_0 at initial time $t = 0$. If the target exhausts his energy, he cannot do anything but staying at his current position.
(4) A payoff of the game is given by the integral of searching resources weighted by the target probability in the entire search space. The searcher plays as a maximizer and the target as a minimizer.

Because the problem is point-symmetric for the origin or the datum point, we adopt distance x from the origin as the coordinate of Euclidean space. We denote the distribution density of searching resources in point $x \in [0, \infty)$ at time t by $h(x, t)$ and the probability density of the target resulting from the target motion by $f(x, t)$. For a searcher's strategy $\boldsymbol{H} = \{h(x,t), x \in [0, \infty), t \in [\tau, T]\}$ and a target's strategy $\boldsymbol{F} = \{f(x,t), x \in [0, \infty), t \in [0, T]\}$, the payoff is given as follows.

$$G(\boldsymbol{H}, \boldsymbol{F}) = \int_\tau^T \int_{X_t} h(x,t) f(x,t) 2\pi x \, dx \, dt, \qquad (1)$$

where X_t indicates the possible area of the target at time t. We have $\int_0^\infty h(x,t) 2\pi x \, dx \leq \rho$, $\tau \leq t \leq T$, as the constraint on searching resources. The probability density $f(x,t) \geq 0$ depends on the target motion and has to satisfy $\int_{X_t} f(x,t) 2\pi x \, dx = 1$, $0 \leq t \leq T$. A sequence of positions of the target taking path p are represented by $p(t)$ and its velocity by

$v(t) \equiv dp(t)/dt$ at time t. On the motion, there are constraints $v(t) \leq V_c$ (maximum speed limit) and $\int_0^T \mu(v(t))dt \leq e_0$ (energy constraint).

The problem is defined on a continuous search space $\mathbf{R}^2$ and has a linear form as a payoff for the players' strategies. However the SAG on a continuous space is usually difficult to solve compared to the SAG on a discrete space. Washburn and Hohzaki [14] did not give an optimal solution but developed the estimation of a lower bound and an upper bound on the value of the game with the help of calculus of variations and optimal control theory. We will outline their results from now.

2.2. Lower bound estimation

Assuming that the target can always make a uniform distribution for his positions within the largest possible area, we can estimate a lower bound on the value of the game. The farthest distance $z(t)$ that the target reaches at time t is given by solving the following problem.

$$z(t) = \max \int_0^t v(\xi)d\xi$$

$$\text{s.t.} \quad 0 \leq v(\xi) \leq V_c, \ 0 \leq \xi \leq t, \ \int_0^t \mu(v(\xi))d\xi \leq e_0.$$

Calculus of variations shows us that a constant-speed motion is an optimal solution. If we overestimate the mobility of the target by the uniform distribution in a circle with the radius of the above farthest distance and use a uniform distribution of searching effort corresponding to the target's motion, we can estimate a lower bound on the value of the game as follows.

$$G_L = \int_\tau^T \frac{\rho}{\pi z(t)^2} dt, \qquad (2)$$

where

$$z(t) = \begin{cases} tV_c, & 0 \leq t \leq e_0/\mu(V_c), \\ t\mu^{-1}(e_0/t), & e_0/\mu(V_c) < t. \end{cases} \qquad (3)$$

For a specified function $\mu(v) = v^2$, the lower bound G_L is calculated in three cases depending on search period $[\tau, T]$.

$$(i) \text{ If } e_0/V_c^2 \leq \tau, \ G_L = \frac{\rho}{\pi e_0} \log(T/\tau), \qquad (4)$$

$$(ii) \text{ if } \tau < e_0/V_c^2 < T, \ G_L = \frac{\rho}{\pi V_c^2}\left(\frac{1}{\tau} - \frac{V_c^2}{e_0}\right) + \frac{\rho}{\pi e_0}\log\frac{TV_c^2}{e_0}, \qquad (5)$$

$$(iii) \text{ if } T \leq e_0/V_c^2, \ G_L = \frac{\rho}{\pi V_c^2}\left(\frac{1}{\tau} - \frac{1}{T}\right). \qquad (6)$$

2.3. Upper bound estimation

If the target takes a feasible motion $y(t)$, which is the target position from origin at time t, he is able to move everywhere within $y(t)$ and generate a uniform distribution of his position in a circle within the radius $y(t)$ at time t. In this case, the expected payoff is $\int_\tau^T \rho/(\pi y(t)^2) dt$, which is just what the target desires to minimize. A rational motion $y(t)$ for the target is given by

$$\min_{\{y(t)\}} \int_0^T I(t)/(\pi y(t)^2) dt$$

$$s.t.\ \dot{y}(t) = v(t),\ 0 \le v(t) \le V_c,\ 0 \le t \le T,\ \int_0^T \mu(v(t)) dt \le e_0, \quad (7)$$

where $I(t) := \{0,\ if\ 0 \le t < \tau;\ 1,\ if\ \tau \le t\}$. Using the optimized motion $y(t)$, we can estimate an upper bound on the value of the game by

$$G_U = \int_\tau^T \frac{\rho}{\pi y(t)^2} dt. \quad (8)$$

For $\mu(v) = v^2$, we apply optimal control theory to solve the problem (7) and obtain an optimal path $y(t)$, its speed $v(t)$ and an upper bound G_U. The results are as follows.

$$y(t) = \begin{cases} Vt, & 0 \le t \le b \\ V\sqrt{b/(T-b)}\sqrt{T(T-b)-(T-t)^2}, & b < t \le T \end{cases} \quad (9)$$

$$v(t) = \begin{cases} V, & 0 \le t \le b \\ V(T-t)\sqrt{b/(T-b)}/\sqrt{T(T-b)-(T-t)^2}, & b < t \le T \end{cases} \quad (10)$$

$$G_U = \frac{\rho}{\pi V^2}\left(\frac{1}{\tau}-\frac{1}{b}\right) + \frac{\rho}{\pi V^2}\frac{1}{2b}\sqrt{\frac{T-b}{T}}\log\frac{\sqrt{T(T-b)}+(T-b)}{\sqrt{T(T-b)}-(T-b)}, \quad (11)$$

where (b, V) is given by the following two cases: (I) $b = \tau$ and $V = \sqrt{l(\tau)}$ in the case of $l(\tau) \le V_c^2$ or (II) $b(> \tau)$ of $l(b) = V_c^2$ and $V = V_c$ in the case of $l(\tau) > V_c^2$ for a function

$$l(x) := \frac{2e_0}{x}\sqrt{\frac{T-x}{T}} \bigg/ \log\frac{\sqrt{T/(T-x)}+1}{\sqrt{T/(T-x)}-1}.$$

3. Simulator for Datum Search Game

To examine the theoretical solution by simulations, we made a simulator, which consists of two personal computers connected to each other via a RS232C cable. The simulator provides a two-dimensional search space, time space from 0 through stopping time T and starting time of the search τ as a player's common search environment. Two PCs are assigned to two operators. They play roles of the target or the searcher out of another's sight. Each of them manipulates a joystick or buttons to move the target or the searcher in a given search time period. The system of the simulator is designed from the standpoint of real search operation so that its functions do not necessarily meet Assumptions (1)–(4) stated in Section 2.1. The biggest difference between the theoretical model and the simulation one occurs on the searcher's side. In the simulation, the searcher has a sensor as searching resources and use it to detect the target. Therefore, the sensor is effective only where the searcher is. However the superiority of the searcher's velocity to the target one would give the searcher a flexible distribution of searching resources in the simulation although the flexibility is not perfect as assumed in the theoretical model.

The console for the use of the target has functions of providing maximum speed V_c, initial energy e_0 and energy consuming rate $\mu(v)$. The display of the target console is shown in Figure 1. In a big circle on the left side of the console, called the position display area, the current position of the player is displayed time by time. The operator can move the target, which is symbolized by a circle with a bar, as shown in Figure 2. The direction of the bar indicates the target heading and its length indicates the target velocity, which are both controllable by the operator. Figure 3 shows us sequential changes of the bar when the operator moves the target to the right. The motion does not change immediately but smoothly. In the right side on the target console, called the information display area, the coordinates of the current position and the residual time are displayed at the top of the area. Additionally, the current speed and residual energy are displayed. The residual amount of moving energy is crucial for the target motion. The display of the residual energy is colored yellow or red according to its warning or critical status, respectively. Below the area, a square window is placed for setting initial parameters for the target, such as maximum speed, initial energy and energy consuming rate function for the target. The bottom window is for watching parameters set for the searcher, such as searching speed, time lag, stopping time and others. To start the simulation, we only need to push a button placed at the bottom.

The searcher console has a similar design to the target console. However it is designed for the operator to not distribute searching resources but move the searcher to search for the target. A position of the searcher is symbolized by a bigger circle with a bar like the target symbol. The direction of the searcher's movement is denoted by the bar. The radius of the circle indicates the effectiveness of the searcher's sensor, called the sweep width. We will explain the sweep width right now.

The linear payoff (1) indicates how well the distributed resources cover the target path or the probability distribution of the target, that is, the effective coverage of searching resources. It is known that for the so-called random search, the probability of detecting the target is given by an exponential function of the effective coverage G,

$$P(G) = 1 - \exp(-\beta\,G)\,, \qquad (12)$$

where β is a parameter indicating the efficiency of unit effective coverage for the detection and is thought to be constant if the search environment remains unchanged. This equation (12) shows the relation between the detection probability and the value of the game in the theory. On the other hand, the detection is a main event which happens when the searcher or his sensor detects the target in the simulation. In search theory, we represent the efficiency of each sensor by a value called sweep width. The event of the detection is stochastic mainly depending on the distance between the sensor and the target if other circumstances are unchanged. For a given target, the dependency on the distance represents the characteristics of the sensor. The sweep width W is an idealized parameter in the sense that the sensor detects the target with certainty if the target passes within the distance W from the sensor, but there is no detection at all otherwise. In the simulation, we set the sweep width for the searcher or his sensor, and simulate the event of the detection when the target and the searcher get closer than W. We calculate the detection probability by transacting the detection events statistically as a result of the simulation.

4. Experiment for a Datum Search Game

Here we are going to compare the results of the simulation with the theory. We take a standard procedure for the simulation as follows. Before the simulation, we set up parameters: time lag τ, stopping time T, a maximum speed of the target V_c, initial energy of the target e_0, energy consuming rate $\mu(v)$, a searcher's speed u and the sweep width of the searching sensor W. These parameters are disclosed to both players as common knowledge and are displayed on the consoles of the players during the simulation. After the simulation begins, the target can move from a center of the position display area by choosing his speed and course under the constraints of maximum speed and energy. He spends energy $\mu(v)$ per second for his current speed v.

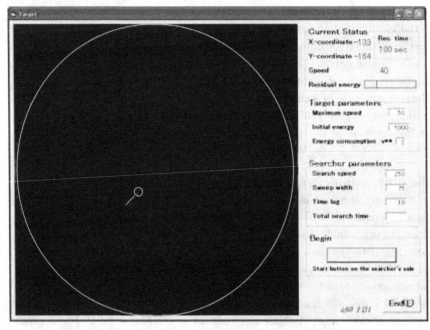

Figure 1. CRT display on target console.

Figure 2. Target.   Figure 3. Change of target heading.

The change of residual energy is also displayed on the console as reference information for the target. If he exhausts initial energy e_0, he is forced to stay where he is since then. The searcher can move from the datum point after time τ but his speed u is unchangeable, which is usually set quite larger than the target, during the simulation. If both players get closer than the sweep width, the detection occurs but they do not know it and continue the search game until the end of the simulation. Except for set-up-parameters, both players are not informed anything about the situation of another player during the simulation. Operators practice repeatedly until they get used to the manipulation of their consoles.

We set parameters as follows: a radius of position display area = 1000 pixels, $\tau = 10$ seconds, $T = 100$ seconds, $V_c = 50$ pixels per second and

$e_0 = 10000$. The setting of energy consuming rate $\mu(v) = v^2$ makes us utilize formulas (4)–(6) and (9)–(11). Related to the searcher, the amount of searching resources $\rho = 1$ is available per second, and a fixed search speed and his sweep width are set to be $u = 250$ pixel per second and $W = 75$ pixels, respectively.

A total of 9 persons joined the experiment and they repeated the simulation 166 times with a time limit $T = 100$. Through the simulation repeated so many times, operators got and learned the feeling about how to play cleverly. This meets a basic but implicit assumption of the game, that is "rational player principle".

(1) Probability of detection

Figure 4 illustrates the cumulative probabilities of detection varying on the axis of time. We calculate the detection probabilities by dividing the total number of detections by the total number of simulations 166. For the theoretical evaluation, we apply G_L or G_U given by Eqs. (4)–(6) or (11) and $\beta = 30000$ to Eq. (12) to obtain a lower bound or an upper bound for the detection probability. As mentioned before, the assumptions of the mathematical model and the functions provided in the simulation are not necessarily corresponding each other. However, curves approximately look consistent in Figure 4. The curves have a typical property that the detection probability is getting saturated as the searching time becomes larger. The decision maker is always asked when he must stop the search in the practical search operation.

(2) Control of target speed

Figure 5 shows two curves with respect to the target speed on the time axis. One is the theoretical result computed by Eq. (10) and the other is the average of speeds that operator took in the simulation. From the observation in the simulation, we can approximately classify the way the operator moved the target into three types: (i) Keep straight in a fixed direction until using up his initial energy so as to go away as far as possible from the datum point. (ii) Move far from the datum point for a while and then come back around the point so as to outwit the searcher. (iii) Move randomly so as not to let the searcher anticipate the target movement correctly.

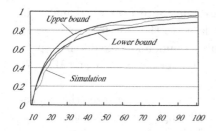

Figure 4. Detection probability.

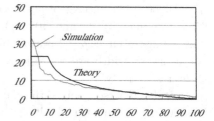

Figure 5. Control of target speed.

There is no way that we know the exact courses the operator took in Figure 5 but we can interpret data about the average speed as follows. The operator takes larger speed at the early time in order to go away as far as possible from the datum point but diminishes his speed gradually so that he can save his energy to maintain his mobility to some extent even at the later time. We can see the similar consideration in the theoretical semi-optimal solution given by Eq. (10) in Fig. 5, although the theory recommends a constant speed at the earlier time. Approximately, we can say that the game-theoretical control of the target speed seems reasonable from the point of view of human psychology. However we must note that the theoretical curve in Fig. 5 signifies the route of the most active target such as that he runs as far as possible from the datum point aiming the minimization of the payoff, e.g. the problem (7). The upper bound G_U is realized by a various types of targets or a mixed strategy of target routes. One of them may be the most active route $y(t)$ but another of them may be more moderate movement which keeps the target staying within $y(t)$. In this sense, we can say that human tends to exhaust his energy more quickly than the game-theoretical recommendation. Thus we can see the difference between the theoretical solution and practical human decision-making as well as their similarity in Fig. 5.

5. Conclusions

This paper deals with a datum search game in a continuous search space. We verify the theoretical solutions of the game through the simulation with human factors. By our simulator, we try to simulate the search operation as realistically as possible and therefore our simulation model does not exactly correspond to the mathematical model of the datum search game. We check the results of the simulation in terms of the probability of detection and a rational control of the target speed and we reach the conclusion that the operators approximately behaves similarly to the theoretical solutions of the game and the operation proceeds according to the way that game theory tells us. These facts may partially depend on the type of the game. Due to a two-person zero-sum game, two operators seem to be competitive enough for us to believe in their incentive. Our results show us that experiments are necessary and useful to verify the rationality and the reasonability of human behavior in search theory as well as experimental economics.

References

1. V.J. Baston and F.A. Bostock, A one-dimensional helicopter-submarine game, *Naval Research Logistics*, **36**, 479–490, (1989).
2. J.M. Danskin, A helicopter versus submarine search game, *Operations Research*, **16**, 509–517, (1968).
3. J.N. Eagle and A.R. Washburn, Cumulative search-evasion games, *Naval Research Logistics*, **38**, 495–510, (1991).
4. A.Y. Garnaev, A remark on a helicopter-submarine game, *Naval Research Logistics*, **40**, 745–753, (1993).
5. J. von Neumann and O. Morgenstern, *Theory of Games and Economic Behavior.* (Princeton University Press, 1944).
6. R. Hohzaki, Search allocation game, *European Journal of Operational Research*, **172**, 101–119, (2006).
7. R. Hohzaki and A.R. Washburn, An approximation for a continuous datum search game with energy constraint, *Journal of the Operations Research Society of Japan*, **46**, 306–318, (2003).
8. K. Kikuta, A search game with traveling cost, *Journal of the Operations Research Society of Japan*, **34**(4), 365–382, (1991).
9. B.O. Koopman, *Search and Screening.* (Pergamon, pp. 221–227, 1980).
10. J.J. Meinardi, A sequentially compounded search game, *Theory of Games: Techniquea and Applications.* (The English Universities Press, London, pp. 285–299, 1964).
11. T. Nakai, A sequential evasion-search game with a goal. *Journal of the Operations Research Society of Japan*, **29**, 113–122, (1986).
12. L.C. Thomas and A.R. Washburn, Dynamic search games, *Operations Research*, **39**, 415–422, (1991).
13. A.R. Washburn, Search-evasion game in a fixed region, *Operations Research*, **28**, 1290–1298, (1980).
14. A.R. Washburn and R. Hohzaki, The diesel submarine flaming datum problem, *Military Operations Research*, **4**, 19–30, (2001).

AN OPTIMAL WAIT POLICY IN TWO DISCRETE TIME QUEUEING SYSTEMS

JUNJI KOYANAGI, DAISUKE NANBA AND HAJIME KAWAI

Department of Social Systems Engineering
Faculty of Engineering, Tottori University
4-101 Koyama Minami, Tottori, 680-8552, Japan
junji@sse.tottori-u.ac.jp,
b02t7037b@edu.tottori-u.ac.jp,
kawai@sse.tottori-u.ac.jp

A system with two discrete time queues is considered. Normal customers arrive at each queue and depart after being served; they cannot choose which queue they join. We consider one special customer who can choose the queue he joins. He also has other options 'wait' and 'leave'. Option 'wait' means that he defers the decision which queue he joins, and option 'leave' means that he leaves the system without being served. If he chooses to join the queue, he stays in the queue until he is served and cost per unit time is incurred corresponding to queueing. Cost for 'wait' is smaller than the waiting cost in the queue. If he chooses to 'leave', cost for 'leave' is incurred and no cost is incurred after that. Our objective is to minimize the total expected cost until he is served or he chooses to 'leave'.

1. Introduction

In a service system, there are several types of services according to the customer type. In a queueing system, FIFO (First-In-First-Out) system is a very common service discipline. However, some queueing systems give special customers priority in service, or give them some choices in service. For these systems, it is important to analyze how beneficial the priorities and the choices are to the special customers. If the special customer has some choices, he would like to know which choice is the best one in various situations.

There are several queueing models that analyze the best choice for the customers. A famous problem of customers' decision is 'shortest queue problem'. Consider a queueing system which contains several indentical queues. Customer can choose which queue he enters when he arrives. It seems natural that his best decision is to join the shortest queue to minimize his waiting time. However, there is a counterexample[1] against this intuition, and there are various papers[2,3] about this topic. In the shortest queue problem, there are no special customers and the decision is made at once.

As a model with multiple decision epochs and special customers, Mandelbaum and Yechiali[4] propose the model where a special customer can choose when to join the queue. In their model, $M/G/1$ queueing system is considered, where decision is made at the end of service of normal customer. The special customer can choose the action among 'join (the queue)', 'wait (in waiting room)' and 'leave (the queue)'. Though the cost for waiting in waiting room is smaller than that for waiting in the queue, the customers that arrive during the time he is in the waiting room are placed before him if and when he decides to join the queue. The optimal policy for this problem has a simple structure, the optimal action changes from 'join' to 'wait' and finally to 'leave' as the system length increases. As a variation of this model, we deal with an optimal policy for a smart customer with two waiting options, short and long stay in a waiting area[5].

In a production system, this type of decision problem may occur. A production system usually produces one kind of item, which is sent to the (queueing) system and the quality of the item is checked at the server. A special kind of item is sometimes sent to the server for the check. Since the production of the special item is fairly rare, it is considered that at most one special item exists in the system. The special item requires to be kept in a special condition, for example, to be kept at low temperature. If the item is not on the line for the check, the condition could be kept with low cost, if the condition is low temperature, in a refrigerator. However, if the item is on the line for the check, more cost is needed to keep the condition, for example, the whole line must be kept low temperature. Thus to minimize the cost for special item, it is important to determine when the special item is put on the line.

If two kinds of items are usually produced, must be checked at two corresponding servers, and the special item could be checked at either server, the problem becomes when and to which server the special item should be sent. The model in this paper deals with the above problem and an extension of the model[4]. We consider a queueing system with two discrete time queues. Normal customers have no choice, they simply enter the queue to which they arrive, while the special customer can choose which queue he enters, defer his decision, or leave the system. By this extension, the state space becomes two dimensional and the special customer has four actions. The problem is formulated as a Markov decision process and the monotone structure of the optimal policy is established.

2. Model

Two discrete time queueing systems are considered in this paper. Each queueing system (QA and QB) has one server and one queue. The normal customers are served in FIFO deciphline at one of the servers. When the system length of QA is i, the customer in service departs from QA with service probability q and one customer arrives at QA with arrival probability a_i at the next time. The arrival probability b_j and the service probability r of QB are defined in a similar manner.

Though the normal customers behave as mentioned above, we consider one special customer who can choose which queue he joins or can leave the system without joining a queue. For the waiting time including the service time in QA or QB, cost c or d per unit time is incurred to him, respectively. If he chooses to leave, cost s is incurred. He departs the system after he is served either at QA or QB, or when he takes the 'leave' action. If there are only three actions mentioned above, the problem is very simple. However, he also has the fourth option that he defers the decision for a unit time. The cost for this action is assumed to be 1 without loss of generality and this action can be taken repeatedly. The objective is to minimize the total expected cost until he decides to leave or finishes his service in either QA or QB.

To analyze this problem, we consider the state space (i, j), the pair of system length of each queue, where i is the system length of QA and j is the system length of QB. For this state space, we prove the monotone property of optimal policy in this paper. Throughout the paper, the word 'increasing' means non-decreasing and 'decreasing' means non-increasing.

With the assumptions made on the arrival and service probabilities, let f_{ik} be the transition probability of system length of QA, then

$$f_{00} = 1 - a_0, \quad f_{01} = a_0, f_{i\,i-1} = q(1 - a_i)$$
$$f_{ii} = (1-q)(1-a_i) + qa_i, \quad f_{i\,i+1} = (1-q)a_i, \; i > 0.$$

The transition probability g_{jl} of QB is

$$g_{00} = 1 - b_0, \quad g_{01} = b_0, g_{j\,j-1} = r(1 - b_j),$$
$$g_{jj} = (1-r)(1-b_j) + rb_j, \quad g_{j\,j+1} = (1-r)b_j, \; j > 0.$$

We consider the following assumption.

Assumption 2.1. *The arrival probabilities a_i and b_i are decreasing in i.*

By Assumption 2.1, transition probabilities f_{ik} and g_{jl} satisfy the following properties.

Lemma 2.1. *The transition probablity f_{ik} has the following properties.*

(1) For all m, $\sum_{k=m}^{\infty} f_{ik}$ is increasing in i.

(2) For all i, $\sum_{k=0}^{\infty} k f_{i+1\,k} - \sum_{k=0}^{\infty} k f_{ik} \leq 1$.

The proof is easy, thus it is omitted. Assumption 2.1 is needed to prove Lemma 2.1(2). The transition probability g_{jl} also has the same properties.

By the theory of Markov decision process [6], we obtain the following optimality equations.

$$W(i,j) = 1 + \sum_{k=0}^{\infty}\sum_{l=0}^{\infty} f_{ik} g_{jl} V(k,l), \tag{1}$$

$$V(i,j) = \min\{W(i,j), c(i+1)/q, d(j+1)/r, s\} \tag{2}$$

The values of $W(i,j)$ and $V(i,j)$ are calculated as limits of the following iteration.

$$V^0(i,j) \equiv 0 \tag{3}$$

$$W^{n+1}(i,j) = 1 + \sum_{k=0}^{\infty}\sum_{l=0}^{\infty} f_{ik} g_{jl} V^n(k,l), \tag{4}$$

$$V^{n+1}(i,j) = \min\{W^{n+1}(i,j), c(i+1)/q, d(j+1)/r, s\} \tag{5}$$

We prove some lemmas to show the structure of optimal policy.

Lemma 2.2. *The functions $V(i,j)$ and $W(i,j)$ are increasing in i and j.*

Proof. We first use induction in n to prove that $W^n(i,j)$ and $V^n(i,j)$, $n \geq 1$, are increasing with respect to i and j.

(1) We note that $V^0(i,j)$ is increasing in i and j.
(2) We prove that if $V^n(i,j)$, $n \geq 0$, is increasing in i and j, then $W^{n+1}(i,j)$ is increasing in i and j. The idea of this proof can be found in the book[7].

We define $\delta^n(0,j) = V^n(0,j)$ and $\delta^n(i,j) = V^n(i,j) - V^n(i-1,j)$, $i \geq 1$. From the induction hypothesis $\delta^n(i,j) \geq 0$, $i \geq 1$.

Then,

$$W^{n+1}(i+1,j) - W^{n+1}(i,j)$$

$$= \sum_{l=0}^{\infty} g_{jl} \sum_{k=0}^{\infty} f_{i+1\,k} \sum_{h=0}^{k} \delta^n(h,l) - \sum_{l=0}^{\infty} g_{jl} \sum_{k=0}^{\infty} f_{ik} \sum_{h=0}^{k} \delta^n(h,l) \quad (6)$$

$$= \sum_{l=0}^{\infty} g_{jl} \sum_{h=0}^{\infty} \sum_{k=h}^{\infty} f_{i+1\,k} \delta^n(h,l) - \sum_{l=0}^{\infty} g_{jl} \sum_{h=0}^{\infty} \sum_{k=h}^{\infty} f_{ik} \delta^n(h,l) \quad (7)$$

$$= \sum_{l=0}^{\infty} g_{jl} \sum_{h=1}^{\infty} \left(\sum_{k=h}^{\infty} f_{i+1\,k} - \sum_{k=h}^{\infty} f_{ik} \right) \delta^n(h,l) \geq 0 \quad (8)$$

For the last inequality, Lemma 2.1 and $\delta^n(i,j) \geq 0$, $i \geq 1$, along with the fact that

$$\sum_{k=0}^{\infty} f_{ik} = 1, \; i \geq 0. \quad (9)$$

Thus $W^{n+1}(i,j)$ is increasing in i, if $V^n(i,j)$ is increasing in i. In a similar way, $W^{n+1}(i,j)$ is increasing in j, if $V^n(i,j)$ is increasing in j.

(3) If $W^{n+1}(i,j)$ is increasing in i and j, it is obvious that $V^{n+1}(i,j)$ is increasing in i and j.

The previous three steps lead to the conclusions that $W^n(i,j)$ and $V^n(i,j)$ are increasing in i and j. Thus $W(i,j)$ and $V(i,j)$, the limits of $W^n(i,j)$ and $V^n(i,j)$, are also increasing in i and j. □

It was shown that $V^n(i,j)$ and $W^n(i,j)$ are increasing functions in i and j. In the following lemma, it is shown that upper bounds for the increments of these functions exist.

Lemma 2.3. *The functions $W(i,j)$ and $V(i,j)$ have the following properties.*

(1) $W(i+1,j) - W(i,j) \leq c/q$, $V(i+1,j) - V(i,j) \leq c/q$.
(2) $W(i,j+1) - W(i,j) \leq d/r$, $V(i,j+1) - V(i,j) \leq d/r$.

Proof. We first use induction with respect to n to prove that $W^n(i,j)$ and $V^n(i,j)$, $n \geq 1$, satisfy

$$W^n(i+1,j) - W^n(i,j) \leq c/q, \; i,j \geq 0. \quad (10)$$
$$V^n(i+1,j) - V^n(i,j) \leq c/q, \; i,j \geq 0. \quad (11)$$

(1) It is obvious that $V^0(i+1,j) - V^0(i,j) \leq c/q$, $i,j \geq 0$.
(2) We use again $\delta^n(0,j) = V^n(0,j)$ and $\delta^n(i,j) = V^n(i,j) - V^n(i-1,j)$, $i \geq 1$. If $V^n(i,j)$ satisfies (11), then $\delta^n(i,j) \leq c/q$ for $j \geq 0$ and $i \geq 1$, which is the inductive assumption. Then,

$$W^{n+1}(i+1,j) - W^{n+1}(i,j)$$

$$= \sum_{l=0}^{\infty} g_{jl} \sum_{h=1}^{\infty} \left(\sum_{k=h}^{\infty} f_{i+1\,k} - \sum_{k=h}^{\infty} f_{ik} \right) \delta^n(h,l) \quad (12)$$

$$\leq \sum_{l=0}^{\infty} g_{jl} \sum_{h=1}^{\infty} \left(\sum_{k=h}^{\infty} f_{i+1\,k} - \sum_{k=h}^{\infty} f_{ik} \right) c/q \quad (13)$$

$$= \sum_{l=0}^{\infty} g_{jl} \sum_{k=1}^{\infty} (k f_{i+1\,k} - k f_{ik}) c/q \quad (14)$$

$$\leq \sum_{l=0}^{\infty} g_{jl} c/q = c/q \quad (15)$$

Here (12) holds in view of (9), the inequality (13) holds by Lemma 2.1(1), and the fact that $\delta^n(i,j) \leq c/q$, $i \geq 1$. In addition, the inequality in (15) holds by Lemma 2.1(2).

(3) If $W^{n+1}(i+1,j) - W^{n+1}(i,j) \leq c/q$, it holds that $V^{n+1}(i+1,j) - V^{n+1}(i,j) \leq c/q$, in view of (5) and the fact that

$$V^{n+1}(i+1,j) - V^{n+1}(i,j)$$
$$\leq \max\{W^{n+1}(i+1,j) - W^{n+1}(i,j), c(i+1)/q - ci/q\} = c/q.$$

Thus, the induction completes the proof of (10) and (11). Taking the limit as n tends to infinity in (10) and (11), we conclude that $W(i,j)$ and $V(i,j)$ satisfy Lemma 2.3(1)

It can be proved, in a similar way, that $W(i,j+1) - W(i,j) \leq d/r$ and $V(i,j+1) - V(i,j) \leq d/r$. □

We denote the action 'go to QA' by 'A', 'go to QB' by 'B', wait by 'W' and 'leave the system' by 'L'. In the next theorem, the structure of optimal policy is explained with this notation.

Theorem 2.1. *The optimal policy has the following structure.*

(1) *If optimal action for (i,j) is 'W', the optimal action is 'W', 'L' or 'B' for (k,j), $k \geq i$.*
(2) *If optimal action for (i,j) is 'L', the optimal action is 'L' for (k,j), $k \geq i$.*

(3) *If optimal action for (i,j) is 'B', the optimal action is 'B' for (k,j), $k \geq i$.*

This means that once optimal action becomes 'W', it never becomes 'A' as i increases, and once optimal action becomes 'L' or 'B', it never changes as i increases. Similar properties hold when j increases.

Proof. If optimal action for (i,j) is 'W', $W(i,j) < c(i+1)/q$, then $W(k,j) \leq W(i,j) + (k-i)c/q < c(k+1)/q$ by Lemma 2.3(1). Thus the optimal action for (k,j) ($k \geq i$) cannot be 'A'.

The proof of (2) and (3) are obvious, because $W(i,j)$ is increasing in i and the costs for action 'B' and 'L' do not change as i increases. □

3. Numerical Example

In this section we show two numerical examples to picture our theorem.

Example 1

(1) Service probablity $q = 0.6$, $r = 0.8$, Waiting cost $c = 2$, $d = 4$, Leaving cost $s = 20$.

(2) Arrival probablity a_i to QA depends on system length i.

$a_0 = 0.95$, $a_1 = 0.75$, $a_2 = 0.75$, $a_3 = 0.75$, $a_4 = 0.55$, $a_5 = 0.45$, $a_6 = 0.35$, $a_7 = 0.35$, $a_8 = 0.25$, $a_9 = 0.25$, $a_{10} = 0.25$, $a_{11} = 0.15$, $a_{12} = 0.15$, $a_{13} = 0.15$, $a_{14} = 0.15$, $a_k = 0$ ($k \geq 15$).

(3) Arrival probablity b_j to QB depends on system length j.

$b_0 = 0.9$, $b_1 = 0.7$, $b_2 = 0.6$, $b_3 = 0.5$, $b_4 = 0.4$, $b_5 = 0.3$, $b_6 = 0.2$, $b_7 = 0.2$, $b_8 = 0.1$, $b_9 = 0.1$, $b_l = 0$ ($l \geq 10$).

The optimal policy for this example is shown in Table 1. The result of Theorem 2.1 seems very natural. However, without Assumption 2.1, the optimal policy could have unnatural structure.

Example 2

In this example, Assumption 2.1 is not satisfied, a_i (b_j) may increase as $i(j)$ increases, and the property of Theorem 2.1 is violated.

(1) Service probablity $q = 0.6$, $r = 0.8$, Waiting cost $c = 2$, $d = 4$, Leaving cost $s = 20$

Table 1. The optimal policy for Example 1.

j																
10	A	A	A	A	A	W	L	L	L	L	L	L	L	L	L	L
	A	A	A	A	A	W	L	L	L	L	L	L	L	L	L	L
	A	A	A	A	A	W	L	L	L	L	L	L	L	L	L	L
	A	A	A	A	A	W	L	L	L	L	L	L	L	L	L	L
	A	A	A	A	A	W	L	L	L	L	L	L	L	L	L	L
5	A	A	A	A	A	W	W	L	L	L	L	L	L	L	L	L
	A	A	A	A	W	W	W	W	W	W	W	W	W	W	W	W
	A	A	A	A	W	W	W	W	W	W	W	W	W	W	W	W
	A	A	A	W	W	W	W	W	W	W	W	W	W	W	W	W
	A	A	W	B	B	B	B	B	B	B	B	B	B	B	B	B
0	A	B	B	B	B	B	B	B	B	B	B	B	B	B	B	B
	0					5					10					15 i

(2) Arrival probablity a_i to QA depends on system length i.

$$a_0 = 0.1, \ a_1 = 0.1, \ a_2 = 0.9, \ a_3 = 0.7, a_4 = 0.2, \ a_5 = 0.1,$$
$$a_6 = 0.1, \ a_7 = 0.1, \ a_8 = 0.8, \ a_9 = 0.7, a_{10} = 0.5, a_{11} = 0.1,$$
$$a_{12} = 0.2, a_{13} = 0.5, a_{14} = 0.1, a_k = 0 \quad (k \geq 15).$$

(3) Arrival probablity b_j to QB depends on system length j.

$$b_0 = 0.1, b_1 = 0.7, b_2 = 0.2, b_3 = 0.1, b_4 = 0.4, b_5 = 0.9,$$
$$b_6 = 0.2, b_7 = 0.2, b_8 = 0.9, b_9 = 0.1, b_l = 0 \quad (l \geq 10).$$

The optimal policy for this example is shown in Table 2. When $j = 6$, the optimal action changes from 'A' to 'W' and again changes to 'A', as i increases.

Table 2. The optimal policy for Example 2.

j																
10	A	W	A	A	W	W	W	L	L	L	L	L	L	L	L	L
	A	W	A	A	W	W	W	L	L	L	L	L	L	L	L	L
	A	W	A	A	W	W	W	L	L	L	L	L	L	L	L	L
	A	W	A	A	W	W	W	L	L	L	L	L	L	L	L	L
	A	W	A	A	W	W	W	L	L	L	L	L	L	L	L	L
5	A	W	A	A	W	W	W	L	L	L	L	L	L	L	L	L
	A	W	A	A	W	W	W	W	W	W	W	W	W	W	W	W
	A	W	A	W	W	W	W	W	W	W	W	W	W	W	W	W
	A	W	A	W	W	W	W	W	W	W	W	W	W	W	W	W
	A	W	W	W	W	W	B	B	B	B	B	B	B	B	B	B
0	A	B	B	B	B	B	B	B	B	B	B	B	B	B	B	B
	0					5					10					15 i

4. Conclusion

This paper dealt with the optimal policy in a queueing system with two discrete time queues. The special customer can choose actions and it is assumed that only one special customer exists in the system. Under some conditions, it is shown that the optimal action changes monotonically as queue lengths increase. It is easy to extend this model to continuous time queueing systems if the arrival and service times have exponential distributions.

References

1. W. Whitt, Deciding which queue to join: some counterexamples, *Operations Research*, **34-1**, 55 (1986).
2. P.K. Johri, Minimizing the number of customers in queuing systems, *European Journal of Operational Research*, **27**, 117 (1986).
3. R.R. Weber, "On the optimal assignment of customers to parallel servers", *Journal of Applied Probability*, **15**, 406 (1978).
4. A. Mandelbaum and U. Yechiali, Optimal entering rules for a customer with wait option at an $M/G/1$ queue, *Management Science*, **29-2**, 174 (1983).
5. J. Koyanagi and H. Kawai, A smart customer in a discrete time queue with two waiting options (in Japanese). *Kôkyûroku in Research Institute for Mathematical Sciences*, **1457**, 194 (2005).
6. S.M. Ross, *Applied Probability Models with Optimization Applications*, Holden-Day, San Francisco, 1970.
7. R.E. Barlow and F. Proschan, *Mathematical Theory of Reliability*, Wiley, New York, 1965.

ANALYSIS OF FINITE OSCILLATING $GI^X/M(m)//N$ QUEUEING SYSTEMS

FÁTIMA FERREIRA

Department of Mathematics and CEMAT
University of Trás-os-Montes e Alto Douro
Quinta dos Prados, 5001-911 Vila Real, Portugal
mmferrei@utad.pt

ANTÓNIO PACHECO

Department of Mathematics and CEMAT
Instituto Superior Técnico, Technical University of Lisbon
Av. Rovisco Pais, 1049-001 Lisboa, Portugal
apacheco@math.ist.utl.pt

HELENA RIBEIRO

Department of Mathematics and CEMAT
School of Technology and Management, Polytechnic Institute of Leiria
Campus 2, Morro do Lena - Alto do Vieiro, 2411-901 Leiria, Portugal
mhcr@estg.ipleiria.pt

In this work we investigate oscillating $GI^X/M(m)//N$ systems, which are queueing systems whose service mechanism reacts to the congestion of the system; namely, the service rates oscillate between two forms according to the evolution of the number of customers in the system. Resorting to Markov chain embedding, we address the time-dependent and limit analysis of the number of customers in the system at batch prearrivals and seen by customers at their arrival to the system. These results are then used to derive the continuous-time limit distribution of the number of customers in the system. In addition, we provide some numerical examples that illustrate our approach.

1. Introduction

We investigate the number of customers in oscillating $GI^X/M(m)//N$ systems, here denoted as $GI^X/M(m) - M(m)//N/(a,b)$ systems. These are queueing systems with finite capacity N, including the customer in service - if any. The customers arrive in batches with independent and identically distributed (i.i.d.) sizes and the batch interarrival times are i.i.d. random variables, independent of the batch sizes. The service is Markovian with service rates moving between two forms according to the evolution of the number of customers in the system, as explained next.

The system oscillates between two operating phases, 1 and 2, which impact the aggregate service rate. When the system is in phase 1 the number of customers moves between 0 and $b-1$, and when it is in phase 2 the number of customers moves between $a+1$ and N, with a and b being two integers such that $0 \leq a < b \leq N$. We call a the *lower barrier* and b the *upper barrier*. The evolution of the phase of the system is as follows. If at time t the system is operating in phase 1, so that the number of customers in the system is smaller than the upper barrier b, then the system remains in phase 1 until the first subsequent epoch at which the number of customers in the system becomes greater or equal to the upper barrier b. At this epoch, the system changes to phase 2 and remains in this phase until the first subsequent epoch at which the number of customers in the system becomes equal to the lower barrier a, at which time the system changes again to phase 1, and so on.

We use the term oscillating system in the sense used in [2, 5, 6]. However, we note that oscillating systems were investigated before in several other works, including [1, 3, 4, 9, 21–24, 26, 27]. Oscillating systems are particular cases of queueing systems with state dependent parameters, in the sense used in the review paper of Dshalalow [7], which includes an extensive list of references on the subject.

A common type of oscillating system is characterized by the service time of a customer having either distribution A_1 or distribution A_2 depending on the phase of the system at the customer arrival being 1 or 2, respectively, as considered, e.g., in [1, 3, 4, 9, 21, 27]. In particular, [3, 4, 27] propose the use of oscillating systems in the analysis of cell-discarding schemes for voice packets in ATM networks, by allowing dropping of low-priority (less significant) bits of information during congestion periods. We note that [20] uses similar models for overload control in message storage buffers such that both the input and service rates or characteristics may depend on the phase of the system.

We derive results for the number of customers in $GI^X/M(m)-M(m)//N/(a,b)$ systems, exploring the Markov regenerative structure of these systems at an appropriate time sequence by means of Markov chain embedding. Specifically, by noting that a $GI^X/M(m) - M(m)//N/(a,b)$ system is a Markov regenerative process (MRGP; see, e.g., [17]) associated to the renewal sequence of batch prearrival epochs (i.e., immediately before the arrival of batches), for the analysis of the number of customers in the system we proceed as in [10] for the analysis of $GI^X/M(m)//N$ systems. Markov chain embedding is used in the analysis of the batch prearrival state process (i.e., the number of customers in the system immediately before the arrival of batches) and of the customer prearrival state process (i.e., the number of customers in the system seen by customers at their arrival to the system).

The characterization of the one-step transition probabilities of the number of customers in the system at batch prearrivals is based on the uniformization of the continuous-time Markov chain associated to the number of customers in the system in-between two consecutive batch arrival epochs. Then, by using the balance equations for equilibrium transition rates or the Markov regenerative structure of the (continuous-time) state process, we express the limit number of customers in the system in continuous-time as a function of the limit distribution of the batch prearrival state process and of the expected sojourn times in states in-between consecutive batch arrivals.

We end this introduction with a brief outline of the paper. In Section 2 we provide some preliminary considerations about the $GI^X/M(m) - M(m)//N/(a,b)$ system and introduce some notation. In sections 3 and 4 we provide the time-dependent analysis of the number of customers in the system, as seen by batches and by individual customers at their arrival to the system, respectively. Capitalizing on these results, in Section 5 we characterize the limit distribution of the number of customers in the system in continuous-time. Finally, in Section 6 some numerical results obtained using the proposed methodology are provided.

2. $GI^X/M(m)//N$ Systems

In $GI^X/M(m) - M(m)//N/(a,b)$ systems, customers arrive in batches according to a general renewal process, and we let A denote the distribution function of a batch interarrival time and λ^{-1} its mean. The batch sizes are i.i.d. random variables with probability function $(f_l)_{l \in \mathbb{N}_+}$, with finite mean $\bar{f}$, where $\mathbb{N}_+$ denotes the set of natural numbers. As regards the customer acceptance policy, we consider the standard policy in queueing systems known in the literature as *partial blocking* (see, e.g., [28]) in which if at arrival of a batch of l customers there are only m, $m < l$, free positions available in the system, then m customers of the batch enter the system and the remaining $l - m$ customers of the batch are blocked. However, our approach can be applied to $GI^X/M(m) - M(m)//N/(a,b)$ systems with stochastic customer acceptance policy, function of the customer batch size and of the number of customers in the system at its arrival.

The service is Markovian with state dependent service rates, which oscillate between two sets according to the evolution of the number of customers in the system, as described in the previous section, and we let μ_c denote the aggregate service rate when the system is in state c. As the system can operate in two different phases, by state c we always mean a pair (c_1, c_2) with c_1 denoting the number of customers in the system and c_2 the phase the system is operating under.

We let $X = (X(t))_{t \in \mathbb{R}}$ denote the continuous-time state process, that is, $X(t) = (X_1(t), X_2(t))$ is the state of the system at time t. We note that X is a MRGP with state space

$$S = \{(c_1, 1) : 0 \leq c_1 \leq b - 1\} \cup \{(c_1, 2) : a + 1 \leq c_1 \leq N\} \quad (1)$$

associated to the renewal sequence $(T_k)_{k \in \mathbb{N}_+}$ of batch arrival epochs, and we consider that the states are ordered in the following way

$$(0, 1) \leq (1, 1) \leq \cdots \leq (b - 1, 1) \leq (a + 1, 2) \leq (a + 2, 2) \leq \cdots \leq (n, 2).$$

Thus, information on the (continuous-time) state process X can and will be obtained from the analysis of the batch prearrival state process X^p (i.e., the discrete time Markov chain (DTMC) embedded immediately before batch arrivals), associated to the Markov renewal sequence $(X_k^p, T_k)_{k \in \mathbb{N}_+}$, and the expected sojourn times in states in-between consecutive batch arrivals.

In the paper, we assume that the batch interarrival time distribution is aperiodic, so that the limit state distribution is also the long-run state distribution (see, e.g., [17, Theorem 9.30]). In case the batch interarrival time distribution is periodic, no limit state distribution exists, but the results derived in the paper remain valid for the long-run state occupancy distribution. We note that, if $a = b - 1$, the system reduces to a $GI^X/M(m)//N$ system (see, e.g., [22]), since its aggregate service rate becomes a function only of the number of customers in the system and, as a result, the *operating phase* of the system may be dropped from consideration. A detailed analysis of $GI^X/M(m)//N$ systems has been carried out in [12].

3. Batch Prearrival State Process

In this section, we analyze the batch prearrival state process X^p, whose limit distribution will be subsequently used to characterize the limit distribution of the continuous-time state process.

X^p is a DTMC with state space S and transition probability matrix

$$P^p = R\Psi, \qquad (2)$$

where: r_{cd} denotes the probability that, given that a batch arrives with the system in state c, the batch postarrival state is d, i.e.,

$$r_{cd} = \sum_{l \in \mathbb{N}_+} f_l \delta_{d_1, \min(c_1+l, N)} = \begin{cases} f_{d_1-c_1} & (c,d) \in \tilde{S} \\ \bar{f}_{N-c_1} & c_1 < d_1 = N \\ 1 & c_1 = d_1 = N \\ 0 & \text{otherwise} \end{cases} \qquad (3)$$

with $\tilde{S} = \{(c,d) : (c_1 < d_1 \leq b-1 \land c_2 = d_2) \lor \max(b-1, c_1) < d_1 < N\}$, where δ denotes the Kronecker delta function, i.e., $\delta_{ij} = 1$ if $i = j$ and $\delta_{ij} = 0$ otherwise, and $\bar{f}_l = \sum_{m \geq l} f_l$; and ψ_{cd} denotes the probability that, given that after the arrival of a batch the system stays in state c, the next batch finds the system in state d at its arrival, i.e.,

$$\psi_{cd} = \mathbf{P}(X(T_{k+1}^-) = d | X(T_k) = c). \qquad (4)$$

Therefore, in order to compute the transition probability matrix P^p, we need to compute the matrix Ψ. In general Ψ does not have a closed form expression. However, as the service is Markovian, in-between consecutive batch arrivals the state process evolves as a continuous-time Markov chain (CTMC) $D = (D(t))_{t \geq 0}$ with state space S and infinitesimal generator matrix Q such that

$$q_{cd} = \begin{cases} \mu_c & d = (c_1 - 1, c_2) \lor (c = (a+1, 2) \land d = (a, 1)) \\ -\mu_c & c_1 > 0 \land c = d \\ 0 & \text{otherwise} \end{cases}.$$

Thus, the stochastic matrix Ψ is equal to the transition probability matrix of D in a batch interarrival time T, independent of D, i.e.,

$$\psi_{cd} = \mathbf{P}(D(T) = d | D(0) = c) = \int_0^\infty \mathbf{P}(D(t) = d | D(0) = c) A(ds) \qquad (5)$$

for $c, d \in S$. Thus, for the computation of the transition probabilities in (5) we let

$$\mu^\star \geq \max_{\{c \in S : c_1 > 0\}} \mu_c$$

denote an upper bound for the aggregate service rates, and consider the embedded uniformized DTMC with uniformization rate $\mu^\star$ associated to

D, which has one-step transition probability matrix (see, e.g., [17] or [25])

$$\hat{P} = I + \frac{Q}{\mu^*} \qquad (6)$$

i.e., $\hat{P} = (\hat{p}_{cd})_{c,d \in S}$ is a stochastic matrix, with $\hat{p}_{(0,1)d} = \delta_{(0,1)d}$, and

$$\hat{p}_{cd} = \begin{cases} \dfrac{\mu_c}{\mu^*} & d = (c_1 - 1, c_2) \vee (c = (a+1, 2) \wedge d = (a, 1)) \\ 1 - \dfrac{\mu_c}{\mu^*} & c = d \\ 0 & \text{otherwise} \end{cases} \qquad (7)$$

for $c \neq (0, 1)$.

By conditioning on the number of events in the interval $[0, T)$ of the uniformizing Poisson process with rate μ^*, independent of the process D, we obtain

$$\Psi = \sum_{l=0}^{\infty} \alpha_l \hat{P}^l \qquad (8)$$

where

$$\alpha_l = \int_0^{\infty} e^{-\mu^* t} \frac{(\mu^* t)^l}{l!} A(dt) \qquad (9)$$

denotes the l-th mixed-Poisson probability with mixing distribution A and rate μ^*. Thus, α_l equals the probability that exactly l renewals take place in the uniformizing Poisson process in-between two consecutive customer batch arrivals to the system.

For computational purposes, the mixed-Poisson probabilities can usually be computed in a fast recursive way (cf., e.g., [10, 18, 29]). In turn, the infinite series in (8) is approximated, at some desired precision, using the properties of mixed-Poisson distributions. Namely, truncating (8) at $L = L(\epsilon)$ such that

$$\sum_{l=0}^{L} \alpha_l \geq 1 - \epsilon \qquad (10)$$

it follows that we may approximate Ψ by

$$\Psi^{(L)} = \sum_{l=0}^{L} \alpha_l \hat{P}^l + \sum_{l=L+1}^{\infty} \alpha_l \hat{P}^L, \qquad (11)$$

satisfying $\| \Psi - \Psi^{(L)} \|_{\infty} \leq 2\epsilon$, with $\| B \|_{\infty}$ denoting the L_{∞}-norm of B[a].

---

[a]The L_{∞}-norm of a matrix $B = (b_{ij})_{i,j \in I}$ is defined as $\| B \|_{\infty} = \max\limits_{i \in I} \sum\limits_{j \in I} |b_{ij}|$.

As a result, using (2) and (11), we conclude that the transition probability matrix of the batch prearrival state process, X^p, is approximated, with an error of at most ϵ, with respect to the L_∞-norm, by the matrix $P^{(L)}$ such that

$$P^{(L)} = R\Psi^{(L)}. \tag{12}$$

Then, letting $\pi = (\pi_c)_{c \in S}$ denote the limit probability vector of the number of customers in the system at batch prearrivals, the vector π is approximated by the stationary vector associated to the matrix $P^{(L)}$, $\pi^{(L)}$, i.e.,

$$\pi^{(L)} = \pi^{(L)} P^{(L)} \text{ and } \pi^{(L)} \mathbf{1} = 1$$

where $\mathbf{1}$ denotes a vector of ones.

4. Customer Prearrival State Process

In this section we will focus on the customer perspective by characterizing the state seen by customers at their arrival to the system, i.e., the customer prearrival state process $X^{cu} = (X_l^{cu})_{l \in \mathbb{N}_+}$, where $X_l^{cu} = (X_{1l}^{cu}, X_{2l}^{cu})$ with X_{1l}^{cu} denoting the number of customers that the l-th customer sees in the system at his arrival after removing the blocked customers that arrive in front of him in his batch, and X_{2l}^{cu} denoting the corresponding phase the system is at that instant. For the sake of simplicity, we will simply say that X_l^{cu} is customer l prearrival state.

It is convenient to let G denote the batch-index process (i.e., we let G_l denote the index of the batch the l-th customer belongs to) and I denote the customer position process (i.e., I_l denotes the position of the l-th customer in his batch) with the positions of the customers in a batch taking successively the values from one up to the size of the batch, starting from the customer at the front of the batch and ending at the one at the rear. Thus, $X_{G_l}^p$ is the state at the prearrival of the l-th customer's batch.

Using the convention that empty sums take the value zero, it follows that $l = \sum_{k=1}^{G_l - 1} X_k + I_l$ and, moreover, for $l \in \mathbb{N}_+$, X_l^{cu} is such that

$$X_{1l}^{cu} = \min(X_{1G_l}^p + I_l - 1, N) \tag{13}$$

and

$$X_{2l}^{cu} = \begin{cases} 1 & X_{2G_l}^p = 1 \wedge X_{1G_l}^p + I_l - 1 < b \\ 2 & \text{otherwise} \end{cases}. \tag{14}$$

Note that $I_l - 1$ is the number of customers from the batch of the l-th customer, G_l, that arrive in front of him and, thus, $X_{G_l}^p + I_l - 1$ is the number of customers seen in the system by customer l at his arrival in case no customers in front of him in his batch are blocked; otherwise the l-th customer sees the buffer full (i.e., with N customers) at his arrival to the system. In turn, a customer sees the system in phase 1 if and only if his batch finds the system in phase 1 at arrival, and the system remains with less than b customers after the incorporation of all customers in front of him in the customer batch.

A careful inspection allows us to assert that the process $Z = (X_G^p, I)$ is an irreducible DTMC with state space $E^* = S \times \mathbb{N}_+$ and one-step transition probabilities

$$p_{(c,j)(d,k)}^* = \begin{cases} 1 - q_j & c = d \wedge k = j+1 \\ q_j \psi_{(\min(c_1+j,N),2)d} & k = 1 \wedge (c_2 = 2 \vee c_1 + j \geq b) \\ q_j \psi_{(c_1+j,1)d} & k = 1 \wedge c_2 = 1 \wedge c_1 + j < b \\ 0 & \text{otherwise} \end{cases} \quad (15)$$

where $q = (q_j)$ denotes the batch size hazard rate function, i.e.,

$$q_j = \frac{\mathbf{P}(X_1 = j)}{\mathbf{P}(X_1 \geq j)} = \frac{f_j}{\sum_{m \geq j} f_m} = 1 - \frac{\bar{f}_{j+1}}{\bar{f}_j}, \quad j \in \mathbb{N}_+.$$

Thus, we can derive the following result that will be useful to obtain long-run and limit results for the state seen by customers at their arrival to the system.

Theorem 4.1. *The DTMC* (X_G^p, I) *is positive recurrent and has stationary probability vector* $\pi^Z = (\pi_{(c,k)}^Z)_{(c,k) \in E^*}$ *given by*

$$\pi_{(c,k)}^Z = \pi_c \frac{\bar{f}_k}{\bar{f}} = \frac{\pi_c}{\bar{f}} \prod_{j=1}^{k-1}(1 - q_j). \quad (16)$$

Proof. The equation (16) gives a stationary distribution of $Z = (X_G^p, I)$ as, in view of (15),

$$\pi_{(c,k)}^Z = \frac{\pi_c}{\bar{f}} \prod_{j=1}^{k-1}(1 - q_j) = \pi_{(c,k-1)}^Z p_{(c,k-1)(c,k)}^* = \sum_{(l,m) \in E^*} \pi_{(l,m)}^Z p_{(l,m)(c,k)}^*$$

for $k > 1$, and $\sum_{(d,m)\in E^*} \pi^Z_{(d,m)} p^*_{(d,m)(c,1)}$ is equal to

$$\sum_{(d,m)\in E^*} \frac{\pi_d}{f} \prod_{j=1}^{m-1}(1-q_j) q_m \big(\psi_{(d_1+m,1)c}\mathbf{1}_{\{d_2=1,d_1+m<b\}}$$
$$+ \psi_{(\min(d_1+m,N),2)c}\mathbf{1}_{\{d_2=2\vee d_1+m\geq b\}}\big)$$

$$= \frac{1}{f}\sum_{d\in S} \pi_d \Big[\sum_{m\in \mathbb{N}_+} f_m \big(\psi_{(d_1+m,1)c}\mathbf{1}_{\{d_2=1,d_1+m<b\}}$$
$$+ \psi_{(\min(d_1+m,N),2)c}\mathbf{1}_{\{d_2=2\vee d_1+m\geq b\}}\big)\Big]$$

$$= \frac{1}{f}\sum_{d\in S} \pi_d p^p_{dc} = \frac{\pi_c}{f} = \pi^Z_{(c,1)}. \qquad \square$$

The previous result establishes that $Z = (X^p_G, I)$ is a DTMC; however, except for the case of single customer arrivals, the state process at customer arrivals, X^{cu}, is not a DTMC. Nevertheless, in view of (13)-(14), the bivariate process (X^p_G, I) may be used in order to derive results for X^{cu}. In particular, in view of (13)-(14) and the total probability law, we have the following result.

Corollary 4.1. *The state seen by customer l at his arrival to the system, X^{cu}_l, $l \in \mathbb{N}_+$, has probability function*

$$p_{X^{cu}_l}(d) = \begin{cases} \sum_{c_1\leq d_1} p_{Z_l}((c_1,1), d_1-c_1+1) & d_1 < b \wedge d_2 = 1 \\ \sum_{a<c_1\leq d_1} p_{Z_l}((c_1,2), d_1-c_1+1) & a < d_1 < b \wedge d_2 = 2 \\ \sum_{c_1<b} p_{Z_l}((c_1,1), d_1-c_1+1) & \\ + \sum_{a<c_1\leq d_1} p_{Z_l}((c_1,2), d_1-c_1+1) & b \leq d_1 < N \\ \sum_{c\in S}\sum_{k\geq N-c_1+1} p_{Z_l}(c,k) & d_1 = N \end{cases} \quad (17)$$

where $(p_{Z_l}(c,j))_{(c,j)\in E^*}$ is the probability function of Z_l, which may be computed by taking into account (15).

In order to derive results for the long-run and limit behavior of the state seen by customers at their arrival to the system, we recall that the long-run fraction and long-run expected fraction of customers that see the system in state d at their arrival to the system are given, respectively, by

$$\lim_{l\to+\infty}\frac{1}{l}\sum_{k=1}^{l}U_{kd} \quad \text{and} \quad \lim_{l\to\infty}\mathrm{E}\left[\frac{1}{l}\sum_{k=1}^{l}U_{kd}\right] = \lim_{l\to+\infty}\frac{1}{l}\sum_{k=1}^{l}p_{X_k^{cu}}(d)$$

where $U_{kd} = 1_{\{X_k^{cu}=d\}}$.

Corollary 4.2. *The long-run fraction and long-run expected fraction of customers that see the system in state d at their arrival to the system are equal and given by*

$$\pi_d^{cu} = \begin{cases} \sum_{c_1 \leq d_1} \pi_{(c_1,1)} \dfrac{\bar{f}_{d_1-c_1+1}}{\bar{f}} & d_2 = 1 \\[2ex] \sum_{a<c_1 \leq d_1} \pi_{(c_1,2)} \dfrac{\bar{f}_{d_1-c_1+1}}{\bar{f}} & a < d_1 < b \wedge d_2 = 2 \\[2ex] \sum_{c_1<b} \pi_{(c_1,1)} \dfrac{\bar{f}_{d_1-c_1+1}}{\bar{f}} + \sum_{a<c_1\leq d_1} \pi_{(c_1,2)} \dfrac{\bar{f}_{d_1-c_1+1}}{\bar{f}} & b \leq d_1 < N \end{cases}$$
(18)

and $\pi_{(N,2)}^{cu} = 1 - \sum_{c \in S\setminus\{(N,2)\}} \pi_c^{cu}$.

Proof. In view of Theorem 4.1, $Z = (X_G^p, I)$ is a positive recurrent DTMC and has stationary distribution $\pi^Z = (\pi_{(c,k)}^Z)_{(c,k)\in E^*}$ given by (16). In addition, the random variables U_{kd} are a bounded function of the positive recurrent DTMC Z, namely

$$U_{kd} = \begin{cases} 1_{\{X_{1G_k}^p+I_k-1=d_1\}}1_{\{X_{2G_k}^p=d_2\}} & d_1 < b \\ 1_{\{X_{1G_k}^p+I_k-1=d_1\}} & b \leq d_1 < N \\ 1_{\{X_{1G_k}^p+I_k-1 \geq N\}} & d_1 = N \end{cases}.$$

Thus, in view of the ergodic theorems for DTMCs (see, e.g., [25, Proposition 2.12.4 and Corollary 2.12.5]), it follows that

$$\lim_{l\to\infty}\frac{1}{l}\sum_{m=1}^{l}U_{md} = \lim_{l\to\infty}\mathrm{E}\left[\frac{1}{l}\sum_{m=1}^{l}U_{md}\right]$$

$$= \begin{cases} \sum_{\{(c,k)\in E^*: c_1+k-1=d_1 \wedge c_2=d_2\}} \pi_{(c,k)}^{cu} & d_1 < b \\ \sum_{\{(c,k)\in E^*: c_1+k-1=d_1\}} \pi_{(c,k)}^{cu} & b \leq d_1 < N \\ \sum_{\{(c,k)\in E^*: c_1+k-1\geq N\}} \pi_{(c,k)}^{cu} & d_1 = N \end{cases} \qquad \square$$

5. Continuous-Time State Process

In this section we characterize the limit distribution of the (continuous-time) state process, X, and its relation to the limit distribution of the batch prearrival state process.

As presented in the next result, an efficient approach is obtained by relating the limit state probability vector $p = (p_d)_{d \in S}$ of the state process X with the limit probability vector of the batch prearrival state process, $\pi = (\pi_c)_{c \in S}$, by means of the balance equations for the limit transition rates. Another alternative, provided in a subsequent remark, is obtained by exploring the Markov regenerative structure of $GI^X/M(m) - M(m)//N/(a,b)$ systems.

Theorem 5.1. *The limit state probability vector $p = (p_c)_{c \in S}$ is such that,*

$$p_c = \frac{1}{\mu_c} \begin{cases} \lambda \sum_{d_1 < c_1} \pi_{(d_1,1)} \bar{f}_{c_1 - d_1} & 1 \leq c_1 \leq a \\ \lambda \sum_{d_1 < b} \pi_{(d_1,1)} \bar{f}_{b - d_1} & c = (a+1, 2) \\ \lambda \sum_{d_1 < c_1} \pi_{(d_1,1)} \bar{f}_{c_1 - d_1} - p_{(a+1,2)} \mu_{(a+1,2)} & c_2 = 1 \wedge a < c_1 < b \\ \lambda \sum_{a < d_1 < c_1} \pi_{(d_1,2)} \bar{f}_{c_1 - d_1} \\ \quad + \lambda \sum_{d_1 < b} \pi_{(d_1,1)} \bar{f}_{\max(c_1, b) - d_1} & c_2 = 2 \wedge c_1 \geq a + 2 \end{cases} \quad (19)$$

for $c \neq (0, 1)$, and

$$p_{(0,1)} = 1 - \sum_{d \in S \setminus \{(0,1)\}} p_d. \quad (20)$$

Proof. In order to prove (19), we will use the balance equations for the limit transition rates, which guarantee that, in the limit, the exit rate from any set of states equals the corresponding entrance rate (see, e.g., [8, Theorem 4.2]).

For that, we let $c_1 \in \{1, \ldots, a\}$ and $F_{c_1} = \{(0,1), (1,1), \ldots, (c_1 - 1, 1)\}$. The entrances into F_{c_1} occur exclusively when a customer leaves the system at a time at which the process X is in state $(c_1, 1)$. Thus, the limit entrance rate into F_{c_1} is given by

$$p_{(c_1,1)} \mu_{(c_1,1)}$$

since the customer service rate in state $(c_1, 1)$ is $\mu_{(c_1,1)}$. On the other side, the exits from F_{c_1} occur when an arrival of a batch with size greater or equal to $c_1 - d_1$ takes place with the process X in state $(d_1, 1) \in F_{c_1}$, and puts the process in a state greater or equal to $(c_1, 1)$. Thus, the limit exit rate from F_{c_1} is

$$\lambda \sum_{d_1 < c_1} \pi_{(d_1,1)} \sum_{l \geq c_1 - d_1} f_l.$$

By equating the limit entrance and exit rates from F_{c_1}, we obtain the first branch of (19).

In order to evaluate $p_{(a+1,2)}$ we now let $G_b = \{(0,1), (1,1), \ldots, (b-1,1)\}$. The entrances into G_b occur exclusively when a customer leaves the system at a time at which the process X is in state $(a+1, 2)$. Thus, the limit entrance rate into G_b is given by

$$p_{(a+1,2)} \mu_{(a+1,2)}$$

since $\mu_{(a+1,2)}$ is the customer service rate in state $(a+1, 2)$. In turn, the exits from G_b occur when an arrival of a batch with size greater or equal to $b - d_1$ takes place with the process X in state $(d_1, 1) \in G_b$, and puts the process in a state greater or equal to $(a+1, 2)$. Thus, the limit exit rate from G_b is

$$\lambda \sum_{d_1 < b} \pi_{(d_1,1)} \sum_{l \geq b - d_1} f_l.$$

By equating the limit entrance and exit rates from G_b, we obtain the second branch of (19).

To proceed, we let $a < c_1 < b$ and $H_{c_1} = \{(0,1), (1,1), \ldots, (c_1 - 1, 1)\}$. In this case, the entrances into H_{c_1} occur whenever a customer leaves the system at a time at which the process X is in state $(a+1, 2)$ or $(c_1, 1)$. Thus, the limit entrance rate into H_{c_1} is given by

$$p_{(a+1,2)} \mu_{(a+1,2)} + p_{(c_1,1)} \mu_{(c_1,1)}$$

since $\mu_{(a+1,2)}$ and $\mu_{(c_1,1)}$ are the customer service rates in states $(a+1, 2)$ and $(c_1, 1)$, respectively. In turn, the exits from H_{c_1} occur when an arrival of a batch with size greater or equal to $c_1 - d_1$ takes place with the process X in state $(d_1, 1) \in H_{c_1}$, and puts the process in a state greater or equal to $(c_1, 1)$. Thus, the limit exit rate from H_{c_1} is

$$\lambda \sum_{d_1 < c_1} \pi_{(d_1,1)} \sum_{l \geq c_1 - d_1} f_l.$$

By equating the limit entrance and exit rates from H_{c_1}, we obtain the third branch of (19).

Finally, we let $c_1 \geq a+2$ and $I_{c_1} = \{(0,1),(1,1),\ldots,(b-1,1),(a+1,2),\ldots,(c_1-1,2)\}$. In this case, the entrances into I_{c_1} occur exclusively when a customer leaves the system at a time at which the process X is in state $(c_1,2)$. Thus, the limit entrance rate into I_{c_1} is given by

$$p_{(c_1,2)}\mu_{(c_1,2)}$$

since $\mu_{(c_1,2)}$ is the customer service rate in state $(c_1,2)$. On the other side, the exits from I_{c_1} occur when an arrival of a batch with size greater or equal to $c_1 - d_1$ takes place with the process X in state $(d_1,2) \in I_{c_1}$ or when an arrival of a batch with size greater or equal to $\max(c_1,b) - d_1$ takes place with the process X in state $(d_1,1)$, which puts the process in a state greater or equal to $(c_1,2)$. Thus, the limit exit rate from I_{c_1} is

$$\lambda \left(\sum_{a < d_1 < c_1} \pi_{(d_1,2)} \sum_{n \geq c_1 - d_1} f_n + \sum_{d_1 < b} \pi_{(d_1,1)} \sum_{l \geq \max(c_1,b) - d_1} f_l \right).$$

Thus, equating the limit entrance and exit rates from I_{c_1}, we obtain the last branch of (19); and (20) follows from the condition $\sum_{c \in S} p_c = 1$. □

Remark 5.1. As X is a MRGP with state space S associated to the renewal sequence $(T_k)_{k \in \mathbb{N}_+}$ of batch prearrival epochs, its limit distribution may also be easily obtained by means of Markov regenerative theory. In fact, using [17, Theorem 9.30] and taking into account that the expected duration of a cycle is the mean batch interarrival time λ^{-1}, independently of the state the system at the beginning of the cycle, i.e.,

$$E[T_{k+1} - T_k | X(T_k^-) = c] = 1/\lambda,$$

for all $c \in S$, it follows that the limit state probability vector p is given by

$$p_d = \lambda \sum_{c \in S} \pi_c \phi_{cd} \qquad (21)$$

where ϕ_{cd} denotes the expected sojourn time of X in state d in-between two consecutive batch arrivals, conditional to the state of the system at the prearrival of the first of these batches being c, i.e.,

$$\phi_{cd} = E\left[\int_{T_k}^{T_{k+1}} 1_{\{X(t)=d\}}\, dt \,\bigg|\, X(T_k^-) = c \right].$$

For the computation of $\Phi = (\phi_{cd})_{c,d \in S}$ we note that

$$\phi_{cd} = \int_0^\infty \int_0^t \sum_{e \in S} r_{ce}\, \mathbf{P}(D(s) = d | D(0) = e)\, ds\, A(du)$$

which, resorting to the uniformization of D with uniformization rate μ^*, leads to

$$\Phi = R \sum_{m \in \mathbb{N}} \bar{\alpha}_m \hat{P}^m \tag{22}$$

where R is the customer acceptance matrix given in (3), $\hat{P}$ is the embedded uniformized transition probability matrix as defined in (6), and

$$\bar{\alpha}_m = \int_{\mathbb{R}_+} \int_0^t e^{-\mu^* s} \frac{(\mu^* s)^m}{m!} \, ds \, A(dt) = \frac{1}{\mu^*}\left(1 - \sum_{l \leq m} \alpha_l\right) \tag{23}$$

denotes the m-th mixed-Poisson expected sojourn time with mixing distribution A and rate μ^* (see [19]).

As, from the properties of the mixed-Poisson expected values, we have that

$$\sum_{l \in \mathbb{N}} \bar{\alpha}_l = 1/\lambda$$

the infinite sum in (22) may be approximated with any desired precision. For that, if we truncate (22) at an order $\bar{N}(\epsilon) \in \mathbb{N}$ such that

$$\sum_{l=0}^{\bar{N}(\epsilon)} \bar{\alpha}_l \geq 1/\lambda - \epsilon \tag{24}$$

then Φ may be approximated by

$$\Phi^* = R \left[\sum_{m=0}^{\bar{N}(\epsilon)} \bar{\alpha}_m \hat{P}^m + \sum_{m=\bar{N}(\epsilon)+1}^{\infty} \bar{\alpha}_m \hat{P}^{\bar{N}(\epsilon)} \right] \tag{25}$$

with $\| \Phi - \Phi^* \|_\infty \leq 2\epsilon$.

6. Numerical Results

In this section, we illustrate the approach proposed in the previous sections to compute limit probabilities and expected values of the number of customers in $GI^X/M(m) - M(m)//N/(a,b)$ systems.

Specifically, to evaluate the influence of the batch size distribution we consider the following batch size distributions with common mean, $\nu \neq 0$: deterministic - the constant ν, $D(\nu)$; geometric with success probability $1/\nu$, $\text{Geo}(1/\nu)$; shifted binomial - a binomial with m trials and success probability $(\nu-1)/m$ added of one unit, $1 + B(m,(\nu-1)/m)$, and Uniform U on the set $\{1, 2, \ldots, 2\nu - 1\}$, $U(1, 2\nu - 1)$. The batch interarrival time distributions in this section have the following parameterizations with common positive mean λ^{-1}: deterministic with value λ^{-1}, $D(\lambda^{-1})$; exponential with rate λ, $M(\lambda)$; Erlang with k phases, $E_k(k\lambda)$; and, Pareto with parameters (β, k) with $\beta > 1$ and $k = (\beta-1)/\beta\lambda$, $P(\beta, (\beta-1)/\beta\lambda)$. We have included the Pareto distribution in the list since its importance in queueing has increased recently with the finding that heavy tailed distributions are adequate to model Internet packet interarrival times [see, e.g., [13–16]].

The results have been computed with MATLAB algorithms with an error of at most $\varepsilon = 10^{-6}$. We have checked the correctness of the implementation of the algorithm proposed by comparing its output either with results obtained from simulations and with results presented for $GI^X/M(m)//N$ systems in [12].

We let $\pi' = (\pi'_i)_{0 \leq i \leq N}$, denote the limit probability vector of the number of customers in the system at batch prearrivals, and $p' = (p'_i)_{0 \leq i \leq N}$, denote the limit probability vector of the number of customer in the system in continuous-time, i.e.,

$$\pi'_i = \begin{cases} \pi_{(i,1)} & 0 \leq i \leq a \\ \pi_{(i,1)} + \pi_{(i,2)} & a+1 \leq i \leq b-1 \\ \pi_{(i,2)} & b \leq i \leq N \end{cases}$$

and

$$p'_i = \begin{cases} p_{(i,1)} & 0 \leq i \leq a \\ p_{(i,1)} + p_{(i,2)} & a+1 \leq i \leq b-1 \\ p_{(i,2)} & b \leq i \leq N \end{cases}.$$

The results illustrate the sensitivity of the limit distributions and mean number of customers in the system at batch prearrivals, in regular $GI^X/M(m)//N$ and $GI^X/M(m) - M(m)//N/(a,b)$ oscillating systems, with respect to the lower and upper barriers, to the interarrival time distribution, and to the batch size distribution. We recall that a regular $GI^X/M(m)//N$ system is the particular case of a $GI^X/M(m) - M(m)//N/(b-1,b)$ oscillating system.

Table 1 presents the behaviour of the limit probability vector of the number of customers in the system at batch prearrivals and in continuous-time, along with the corresponding mean and standard deviation, for $D(2)^{Geo(1/2)}/M(2/3) - M(10/9)//20/(a,b)$ systems, for different values of the barriers (a,b). The service rates considered are of the form $\mu_{(i,1)} = 2/3$, $i = 0, 1, \ldots, a$, and $\mu_{(i,2)} = 10/9$, $i = a+1, a+2, \ldots, N$. As can be seen,

Table 1. Limit probability vector, mean and standard deviation of the number of customers in the system at batch prearrivals and in continuous-time in $D(2)^{Geo(1/2)}/M(2/3) - M(10/9)//20/(a,b)$ systems.

k	$D(2)^{Geo(1/2)}/M(2/3) - M(10/9)/20//(a,b)$					
	$(a,b) = (0,1)$		$(a,b) = (5,15)$		$(a,b) = (14,15)$	
	$100\pi'_k$	$100p'_k$	$100\pi'_k$	$100p'_k$	$100\pi'_k$	$100p'_k$
0	21.9985	11.3922	0.7528	0.3288	0.0603	0.0255
1	8.8752	9.8993	0.8070	0.5646	0.0644	0.0452
2	8.0183	8.9435	1.2688	0.8875	0.1013	0.0709
3	7.2441	8.0800	1.9939	1.3953	0.1590	0.1114
4	6.5447	7.2998	3.1174	2.1931	0.2498	0.1749
5	5.9128	6.5950	4.7734	3.4346	0.3926	0.2748
6	5.3419	5.9582	5.5506	4.5815	0.6172	0.4319
7	4.8261	5.3830	6.3289	5.4738	0.9701	0.6788
8	4.3600	4.8632	6.9933	6.2528	1.5250	1.0670
9	3.9391	4.3936	7.5330	6.9138	2.3977	1.6772
10	3.5588	3.9694	7.9250	7.4441	3.7709	2.6369
11	3.2151	3.5861	8.1231	7.8178	5.9211	4.1466
12	2.9047	3.2399	8.0410	7.9875	9.1752	6.5141
13	2.6244	2.9270	7.5692	7.8733	13.5361	10.1385
14	2.3717	2.6445	6.7214	7.3788	17.7060	15.2213
15	2.1435	2.3895	5.8359	6.5055	11.2439	12.5341
16	1.9288	2.1593	5.2515	5.8789	10.1180	11.3268
17	1.6901	1.9476	4.6017	5.3026	8.8659	10.2165
18	1.3501	1.7344	3.6759	4.7220	7.0824	9.0979
19	0.8496	1.4747	2.3132	4.0152	4.4567	7.7360
20	0.3024	1.1197	0.8234	3.0485	1.5864	5.8735
Mean	5.4992	6.4431	10.8098	11.7651	14.0993	14.9976
St.Dev.	5.2560	5.3852	4.4138	4.4795	2.9856	3.0557

the number of customers in the system strongly depends on the barriers considered.

In Figure 1 we plot the mean number of customers in $GI^{D(3)}/M(0.9) - M(1.1)//20/(a,b)$ systems with service rate equal to 0.9 in phase 1 and 1.1 in phase 2, for several different batch interarrival time distributions with mean 3, and as a function of the barriers. In the graphics of the figure, we may observe that the mean number of customers in the sys-

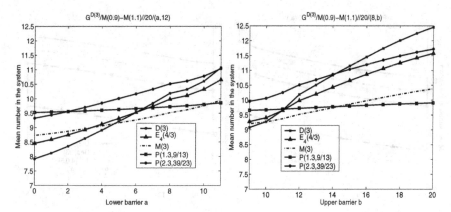

Figure 1. Mean number of customers in $G^{D(3)}/M(0.9) - M(1.1)//20/(a,b)$ systems, with batch arrival rate 1/3, as a function of the lower and of the upper barriers.

tem is sensitive to the values of the lower and upper barriers but also to the interarrival time distribution. Among the studied systems, the system with deterministic batch interarrival times reveals to be the one in which the mean number of customers in the system exhibit higher sensitivity to the values of the barriers. In opposition, in systems with heavy-tailed Pareto batch interarrival time distribution (here corresponding to a $P(1.3, 9/13)^{D(3)}/M(0.9) - M(1.1)//20/(a,b)$ system), this performance measure is practically insensitive to the values of the barriers. This may be explained in part by the tendency that a heavy-tailed Pareto renewal arrival process has to produce bunches of short interarrival times (so that the system rapidly tends to overload and operate in phase 2, independently of the values of the upper barrier b), as a compensation for the observation from time to time of very long interarrival times (in which the system tends to empty and thus to operate in phase 1, independently of the values of the lower barrier a). This is in agreement with the findings of [22] and [11] for customer loss probabilities.

Figure 2 shows how the mean number of customers in the system varies as the lower and upper barriers change for $P(1.3, 9/13)^X/M(0.9) - M(1.1)/1/20$ systems, with mean batch size equal to 3 and batch interarrival rate 1/3. The figure shows that the batch size distribution impacts the mean number of customers in the system.

Finally, to illustrate the impact of the mean batch size on the number of customers in the system, we have considered two batch size distributions: shifted binomial and geometric. Figure 3 shows how the mean number of

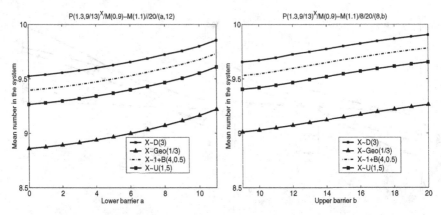

Figure 2. Mean number of customers in $P(1.3, 9/13)^X/M(0.9) - M(1.1)//20/(a,b)$ systems, with batch arrival rate 1/3, as a function of the lower and of the upper barriers.

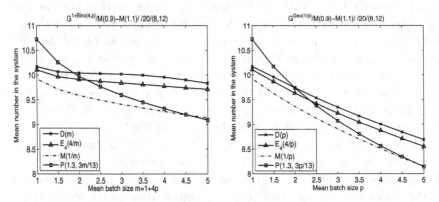

Figure 3. Mean number of customers in $GI^X/M(0.9) - M(1.1)//20/(8, 12)$ systems with unit customer arrival rate, as a function of the mean batch size.

customers in the system evolves as the mean batch size increases, with the (individual) customer arrival rate ($\lambda \bar{f}$) kept fixed with value one.

As expected, across the considered systems with unit customer arrival rate, the mean number of customers in the system shows a tendency to decrease as the mean batch size increases. This came either from the fact that an increase of the batch size leads to an increase of customer blocking and also will put the system working longer in phase 2. This effects are even stronger for the system with Pareto interarrival times due to the heavy-tailed nature of the distribution.

References

1. E. Bahary and P. Kolesar. Multilevel bulk service queues. *Operations Research*, Vol. 20, pp. 406–420, 1972.
2. M. Bratiychuk and A. Chydzinski. On the ergodic distribution of oscillating queueing systems. *Journal of Applied Mathematics and Stochastic Analysis*, Vol. 16, No. 4, pp. 311–326, 2003.
3. B. D. Choi and D. I. Choi. Queueing system with queue length dependent service times and its application to cell discarding scheme in ATM networks. *IEE Proceedings-Communications*, Vol. 143, No. 1, pp. 5–11, 1996.
4. D. I. Choi, C. Knessl, and C. Tier. A queueing system with queue length dependent service times, with applications to cell discarding in ATM networks. *Journal of Applied Mathematics and Stochastic Analysis*, Vol. 12, No. 1, pp. 35–62, 1999.
5. A. Chydzinski. The $M/G - G/1$ oscillating queueing system. *Queueing Systems*, Vol. 42, No. 3, pp. 255–268, 2002.
6. A. Chydzinski. The oscillating queue with finite buffer. *Performance Evaluation*, Vol. 57, No. 3, pp. 341–355, 2004.
7. J. H. Dshalalow. Queueing systems with state dependent parameters. In *Frontiers in Queueing: Models and Applications in Science and Engineering*, pp. 61–116. CRC, Boca Raton, FL, 1997.
8. M. El-Taha and S. Stidham, Jr. *Sample-Path Analysis of Queueing Systems*. Kluwer, Boston, MA, 1999.
9. A. Federgruen and H. C. Tijms. Computation of the stationary distribution of the queue size in an $M/G/1$ queueing system with variable service rate. *Journal of Applied Probability*, Vol. 17, No. 2, pp. 515–522, 1980.
10. F. Ferreira. Embedding, Uniformization and Stochastic Ordering in the Analysis of Level-Crossing Times and $GI^X/M(n)//c$ Systems. PhD thesis, Instituto Superior Técnico, Technical University of Lisbon, Portugal, 2007.
11. F. Ferreira and A. Pacheco. Analysis of $GI/M/s/c$ queues using uniformization. *Computers and Mathematics with Applications*, Vol. 51, pp. 291–304, 2006.
12. F. Ferreira and A. Pacheco. Analysis of $GI^X/M(n)//c$ systems with stochastic customer acceptance policy. *Queueing Systems*, 2008. In print.
13. M. J. Fischer, D. M. Masi, D. Gross, J. Shortle, and P. H. Brill. Development of procedures to analyze queueing models with heavy-tailed interarrival and service times. In *NSF Design, Service, and Manufacturing Grantees and Research Conference, January, 3-6, 2005*, 2005.
14. J. Gordon. Pareto process as a model of self-similar packet traffic. In *Global Telecommunications Conference, 1995 - GLOBECOM '95 3*, pp. 2232–2236, 1995.
15. C. M. Harris, P. H. Brill, and M. J. Fischer. Internet-type queues with power-tailed interarrival times and computational methods for their analysis. *INFORMS Journal on Computing*, Vol. 12, No. 4, pp. 261–271, 2000.

16. Y. Koh and K. Kim. Evaluation of steady-state probability of Pareto/$M/1/K$ experiencing tail-raising effect. *Lecture Notes in Computer Science*, Vol. 2720, pp. 561–570, 2003.
17. V. G. Kulkarni. *Modeling and Analysis of Stochastic Systems*. Chapman and Hall, London, 1995.
18. M. Kwiatkowska, G. Norman, and A. Pacheco. Model checking CSL until formulae with random time bounds. *Lecture Notes in Computer Science*, Vol. 2399, pp. 152–168, 2002.
19. M. Kwiatkowska, G. Norman, and A. Pacheco. Model checking expected time and expected reward formulae with random time bounds. *Computers and Mathematics with Applications*, Vol. 51, No. 2, pp. 305–316, 2006.
20. San-Qi Li. Overload control in a finite message storage buffer. *IEEE/ACM Transactions Communications*, Vol. 37, No. 12, pp. 1330–1337, 1989.
21. J. Loris-Teghem. Hysteretic control of an $M/G/1$ queueing system with two service time distributions and removable server. In *Point Processes and Queuing Problems*, Vol. 24 of *Colloq. Math. Soc. János Bolyai*, pp. 291–305. North-Holland, Amsterdam, 1981.
22. A. Pacheco and H. Ribeiro. Consecutive customer losses in oscillating $GI^X/M//n$ systems with state dependent services rates. *Annals of Operations Research*, 2008. In print.
23. M. F. Ramalhoto. Some inventory control concepts in the control of queues. In W. C. Vogt and M. H. Mickle, editors, *Modelling and Simulation*, Vol. 22, pp. 639–647. University of Pittsburg Press, 1991.
24. M. F. Ramalhoto and R. Syski. Queueing and quality service. Investigacao Operacional, Vol. 16, pp. 155–172, 1996.
25. S. Resnick. *Adventures in Stochastic Processes*. Birkhäuser, Boston, MA, 1992.
26. H. Ribeiro. *Customer Loss Probabilities and Other Performance Measures of Regular and Oscillating Queueing Systems*. PhD thesis, Instituto Superior Técnico, Technical University of Lisbon, Portugal, 2007.
27. K. Sriram, R. S. McKinney, and M. H. Sherif. Voice packetization and compression in broadband ATM networks. *IEEE Journal on Selected Areas in Communications*, Vol. 9, No. 3, pp. 294–304, 1991.
28. P. Vijaya Laxmi and U. C. Gupta. Analysis of finite-buffer multi-server queues with group arrivals: $GI^X/M/c/N$. *Queueing Systems*, Vol. 36, No. 1-3, pp. 125–140, 2000.
29. G. E. Willmot. On recursive evaluation of mixed-Poisson probabilities and related quantities. *Scandinavian Actuarial Journal*, Vol. 2, pp. 114–133, 1993.

A CONTINUOUS-TIME SEAT ALLOCATION MODEL WITH UP-DOWN RESETS

KIMITOSHI SATO
Nanzan University,
Seirei-cho, Seto-shi, 489-0863, Japan,
d07mm002@nanzan-u.ac.jp

KATSUSHIGE SAWAKI
Nanzan University,
Seirei-cho, Seto-shi, 489-0863, Japan
sawaki@nanzan-u.ac.jp

In this paper we consider an airline seat inventory model, in which the planning horizon is continuous time, and the demands for low and high fares are independent random variables, respectively. An initial booking limit for the low fare demands can be revised to reset upward or downward, depending on whether the amount of high fare demands is large enough or not after the certain period of time. After developing a general framework of the continuous-time seat inventory model, we derive the expected revenue function under the fore-mentioned booking policy and explore some analytical properties of optimal booking policies. This model takes into account the Japanese practices in the airline industry. Our model differs from existing literatures in a sense that booking policies are allowed for upward or downward reset. We also provide some numerical examples to illustrate the value function and optimal booking policies.

1. Introduction

Most static revenue management models attempt to maximize the expected revenue in a single-period seat inventory model, when the booking limit for low fare demands is fixed (Sawaki [1]). In these models, once the booking procedure has been stopped, it is never reopened. But there are several papers discussing the dynamic airline seat inventory control (See Talluri and Ryzin [2] and Brumelle and Walczak [3]). Most of them are related to discrete time.

In this paper we consider a seat inventory model which differs from the existing literatures in a sense that the reservation-event can take place continuously. Thus, we treat with demand processes as the Brownian motions because it is easy to treat it mathematically and it is possible to regard the negative demand as the cancellation of the demand. There are some papers that treat with demand processes as the Brownian motions like Chao [4] and Vickson [5]. We assume two fare classes for a set of identical seats. Furthermore, booking limits for the fare class can be reset to upward or downward, depending on whether the amount of high fare demands is large enough or not after a certain period of time. Some passengers book for the high fare ticket, because the low fare ticket may be possibly canceled due to reset downward. Thus, the high fare demands may appear in an early period. If the number of the high fare demands is larger than a certain level prespecified, then the booking limit for the low fare demands should be revised to the level upward. If the number of the high fare demand is smaller than the specific level, then it should be reset downward to accept more reservations of low fare demands. These reservation policies are quite practical because they include a learning process reflecting on the actual booking process. It is found that the results obtained here are in closer agreement with actual airline practices in Japan.

In addition, we develop the reset model to incorporate cancellations by considering callable property. Gallego et al. [6] introduces callable products into a finite-capacity two-period booking process where low-fare customers book first. They analyze the effects of offering callable products on the revenue function and show that they can provide a riskless source of additional revenue to the capacity provider. A three-period model with overbooking and no-show was recently studied in Gallego et al. [6] as an extension of Lee [7]. Besides, they extend the model to overbooking, allowing no-shows and charging a denied boarding cost. Our model does not assume the low-before-high fare booking arrival pattern, so called the early bird assumption. Brumelle et al. [8] considers two possible dependent booking classes and derives optimal conditions under mild assumptions on the dependence structure.

In section 2, we develop a general framework of the continuous-time seat inventory model and then derive the expected revenue function under such booking policies. In section 3, we develop a reset model to incorporate cancellations. First, we extend the reset model in section 2 to the model that the airline pays all low fare passengers whose tickets are recalled in reset downward. Next, we consider the airline seat allocation between high and low fares with a callable property. In section 4, we explore some properties of value functions by using numerical examples.

2. An Airline Seat Inventory Control with the Booking Limits of Upward and Downward Resets

In this section, we present a general framework of the continuous-time airline seat inventory control model. We also consider a special case and the concept of spill rates.

2.1. *The model and sequence of operations*

Our model possesses the characteristic that an initial booking limit for the low fare demands can be revised to reset upward or downward, depending on whether the initial amount of high fare demands is large enough or not at the period specified in advance before the departure. Assume that there are two classes of passengers for low and high fares. Also, assume that demands for the two classes come in a way of a mixture, that is, early birds are not assumed, and low and high demands are independent to each other. Let the planning horizon be a finite closed interval of continuous time $[0, T]$. Let X_t be the random variable of the number of high fare demands at time t and Y_t the number of low fare demands at time t, where uncertainty comes from Brownian motions, respectively, which satisfy

$$dX_t = \mu_1 dt + \sigma_1 dZ_1(t), \quad X_0 = 0 \tag{1}$$

and

$$dY_t = \mu_2 dt + \sigma_2 dZ_2(t), \quad Y_0 = 0 \tag{2}$$

where $Z_1(\cdot)$ and $Z_2(\cdot)$ are independent and standard Brownian motions with mean 0 and variance t. Note that $E[X_t] = \mu_1 t$ and $Var[X_t] = \sigma_1^2 t$. Negative values of demands can be treated as cancellations.

Define t_0 the reset time at which the booking limit for the low fare demands can be revised to reset. Define $\overline{X}^{t_2-t_1}$ the cumulative demands for high fare from time t_1 up to t_2,

$$\overline{X}^{t_2-t_1} = \int_{t_1}^{t_2} X_s ds, \quad 0 \leq t_1 < t_2 \leq T, \tag{3}$$

and similarly, for low fare, as

$$\overline{Y}^{t_2-t_1} = \int_{t_1}^{t_2} Y_s ds, \quad 0 \leq t_1 < t_2 \leq T. \tag{4}$$

Note that at the reset time t_0 we have the cumulative demands $\overline{X}^{t_0-0} = \overline{X}^{t_0}$. It is easy to see from Ross [9] that the integrated Brownian motion can be given by the following Lemma.

Lemma 1. *The cumulative demand from time t_1 up to t_2, $\overline{X}^{t_2-t_1}$, follows a normal distribution with the mean $\mu_1(t_2^2 - t_1^2)/2$ and the variance $\sigma_1^2(t_2^3 - 3t_1^2 t_2 + 2t_1^3)/3$.*

Remark 2. Note that $\overline{X}^{t_0}$ and $\overline{X}^{T-t_0}$ are not independent to each other.

Furthermore, we use the following notations:

C = the total number of seats available,
L = the initial booking limit for low fare demands before the reset time t_0,
p = the high fare,
q = the low fare, $p > q$.
$a \wedge b = \min(a, b)$, $a^+ = \max(a, 0)$.

The class of booking policies is restricted within the narrow limits satisfying the following procedure:

(i) Choose L, α, β, and γ satisfying $0 \leq \beta L \leq C$, $\alpha \leq 1$ and $\gamma \leq 1$.
(ii) Observe the cumulative booking requests from the high fare $\overline{X}^{t_0}$ at the reset time t_0, $0 \leq t_0 \leq T$.

  (a) If $\overline{X}^{t_0} \geq \gamma(C - L)$, then the initial booking limit L should be reset downward to the level αL for $\alpha \leq 1$, that is, more high fare demands are protected after time t_0 to up to T.
  (b) If $\overline{X}^{t_0} < \gamma(C - L)$, then the initial booking limit L should be reset upward to the level βL for $\beta \geq 1$, that is, the seats for low fare demands are available more than the initial seats allocated before t_0. If we choose the parameter β too large, then the number of seats allowed to book for high fare passengers in $[t_0, T]$ may likely be lower than $\overline{X}^{t_0}$. Thus, we assume that the parameter β should be chosen so as to satisfy $C - \beta L \geq \gamma(C - L)$. Hence, the range of β is $1 \leq \beta \leq \{C - \gamma(C - L)\}/L$.

Note that the booking policy can be characterized by the number L and parameters α, β and γ.

Let $v_1(L, \overline{X}, \overline{Y})$ be the total revenue from the booking policy if $\overline{X}^{t_0} < \gamma(C - L)$ and $v_2(L, \overline{X}, \overline{Y})$ the total revenue if $\overline{X}^{t_0} \geq \gamma(C - L)$. v_1 and v_2 consist of the revenues during the time periods $[0, t_0]$ and $[t_0, T]$, respectively. Assume that we are allowed to ignore the cost of denied boarding of low fare customers due to the downward reset. That is, the airlines refunds low fare price q for low fare passengers. This assumption should be relaxed in section 3. In this section, it is assumed that the initial bookings of low fare demands are treated as tentative bookings. Those bookings are confirmed at time t_0. Also, we assume that $P(\overline{X}^{t_0} < C - L) = 1$, that is, the high fare demands do not exceed the protected level in $[0, t_0]$.

Then, the revenue in case (a) is given by

$$v_1(L, \overline{X}, \overline{Y}) = q\min\{L \wedge \overline{Y}^{t_0} + \overline{Y}^{T-t_0}, \beta L\}$$
$$+ p\min\{\overline{X}^{t_0} + \overline{X}^{T-t_0}, C - \beta L \wedge (\overline{Y}^{t_0} \wedge L + \overline{Y}^{T-t_0})\} \quad (5)$$

and similarly in case (b),

$$v_2(L, \overline{X}, \overline{Y}) = q\min\{\alpha L \wedge \overline{Y}^{t_0} + \overline{Y}^{T-t_0}, \alpha L\}$$
$$+ p\min\{\overline{X}^{t_0} + \overline{X}^{T-t_0}, C - \alpha L \wedge (\alpha L \wedge \overline{Y}^{t_0} + \overline{Y}^{T-t_0})\}. \quad (6)$$

Defining $V(L)$ the expected total revenue over the time interval $[0, T]$ when L seats are initially allocated to the low fare demand, $V(L)$ can be written as

$$V(L) = E\left[v_1(L, \overline{X}, \overline{Y})1_{\{\overline{X}^{t_0} < \gamma(C-L)\}} + v_2(L, \overline{X}, \overline{Y})1_{\{\overline{X}^{t_0} \geq \gamma(C-L)\}}\right]. \quad (7)$$

Our objective is to find the optimal booking limit L^* maximizing $V(L)$. At this moment it is difficult to show that $V(L)$ is concave or unimodal in L. However, for either special case of $t_0 = 0$ or $t_0 = T$ under $\alpha = \beta = \gamma = 1$, we can easily an obtain explicit form of $V(L)$: from equation (7),

$$\lim_{t_0 \to 0} V(L) = \lim_{t_0 \to T} V(L) = E\left[q\min(\overline{Y}^T, L) + p\min(\overline{X}^T, C - L \wedge \overline{Y}^T)\right]$$

which is concave in L. As above, if we consider that the demands $\overline{X}^T$ and $\overline{Y}^T$ are the total demand in the reservation period, and we assume the low fare demands occurs earlier than high-fare demands, then the model can be reduced to the traditional model.

Instead of exploring some conditions for concavity of V in L, we consider some special cases when specific assumptions on demands are imposed in the following subsection.

2.2. A case of low fare demand large enough

Suppose that there is a sufficiently large amount of low fare demands which are much bigger than the booking limit for low fare demands. This allows us to ignore the probability distribution function of low fare demand. Let

$F_{\overline{X}^{t_0}}$ and $F_{\overline{X}^{T-t_0}}$ be the probability distribution of cumulative high fare demands at the reset time t_0 and the departure time T, respectively. In the case of low fare demand large enough, we may delete $\overline{Y}$ from the arguments of the functions v_1 and v_2 which can be rewritten as

$$v_1(L,\overline{X}) = q\beta L + p\min\{\overline{X}^{t_0} + \overline{X}^{T-t_0}, C - \beta L\} \tag{8}$$

and

$$v_2(L,\overline{X}) = q\alpha L + p\min\{\overline{X}^{t_0} + \overline{X}^{T-t_0}, C - \alpha L\}. \tag{9}$$

We have the expected revenue $V(L)$ as follows;

$$V(L) = E\left[v_1(L,\overline{X})1_{\{\overline{X}^{t_0} < \gamma(C-L)\}} + v_2(L,\overline{X})1_{\{\overline{X}^{t_0} \geq \gamma(C-L)\}}\right]$$
$$= pC - \int_0^{\gamma(C-L)} T_{x_1}(\beta L)dF_{\overline{X}^{t_0}}(x_1) - \int_{\gamma(C-L)}^{\infty} T_{x_1}(\alpha L)dF_{\overline{X}^{t_0}}(x_1), \tag{10}$$

where

$$T_{x_1}(s) = (p-q)s + p\int_0^{C-s-x_1} F_{\overline{X}^{T-t_0}}(x_2)dx_2. \tag{11}$$

Remark 3. Both $E[v_1(L,\overline{X})]$ and $E[v_2(L,\overline{X})]$ are concave in L but $V(L)$ is not necessarily concave in L.

We now investigate the booking limit L^* maximizing the expected revenue $V(L)$. Differentiating $V(L)$ with respect to L, we obtain

$$\gamma f_{\overline{X}^{t_0}}(\gamma(C-L))\{T_{\gamma(C-L)}(\alpha L) - T_{\gamma(C-L)}(\beta L)\}$$
$$= \beta p \Pr[\overline{X}^{t_0} + \overline{X}^{T-t_0} > C - \beta L \mid \overline{X}^{t_0} \leq \gamma(C-L)]$$
$$+ \alpha p \Pr[\overline{X}^{t_0} + \overline{X}^{T-t_0} > C - \alpha L \mid \overline{X}^{t_0} > \gamma(C-L)]$$
$$- q\{\beta F_{\overline{X}^{t_0}}(\gamma(C-L)) + \alpha \overline{F}_{\overline{X}^{t_0}}(\gamma(C-L))\}. \tag{12}$$

The optimal booking limit L^* satisfying equation (12) may not be unique. Note that the model can be reduced to the traditional model when we assume $t_0 = T$ and $\alpha = \beta = \gamma = 1$. In this case, the optimal booking limit $\hat{L}$ can be rewritten as

$$\hat{L} = \max\left\{0 \leq L \leq C : \Pr[\overline{X}^T \leq C - L] \geq \frac{q}{p}\right\}.$$

2.3. Spill rates

There are two possible interpretations of the term "spill rate" in the airline context. The first is that the spill rate is the expected proportion of flights on which some high fare reservations must be refused because of low fare bookings. The second is that the spill rate is the expected proportion of high fare reservations that must be refused out of the total number of such reservations. It seems that the second be more meaningful since it relates more closely to the amount of high fare revenue lost. The proportion of flights refusing high fare reservations, called the *flight spill rate*, can be expressed as

$$S_1 = \Pr\{\overline{X}^{t_0} + \overline{X}^{T-t_0} + \beta L > C, \ \overline{X}^{t_0} \leq \gamma(C-L)\}$$
$$+ \Pr\{\overline{X}^{t_0} + \overline{X}^{T-t_0} + \alpha L > C, \ \overline{X}^{t_0} > \gamma(C-L)\}$$
$$= \int_0^{\gamma(C-L)} \overline{F}_{\overline{X}^{T-t_0}}(C - \beta L - x_1) dF_{\overline{X}^{t_0}}(x_1)$$
$$+ \int_{\gamma(C-L)}^{\infty} \overline{F}_{\overline{X}^{T-t_0}}(C - \alpha L - x_1) dF_{\overline{X}^{t_0}}(x_1). \tag{13}$$

Next, we have the expected proportion of high fare reservations refused S_2 called the *passenger spill rate* ;

$$S_2 = \frac{1}{E[\overline{X}^{t_0} + \overline{X}^{T-t_0}]}\{$$
$$+ E[\overline{X}^{t_0} + \overline{X}^{T-t_0} - (C - \beta L) | \ \overline{X}^{t_0} \leq \gamma(C-L), \overline{X}^{t_0} + \overline{X}^{T-t_0} > C - \beta L]$$
$$+ E[\overline{X}^{t_0} + \overline{X}^{T-t_0} - (C - \alpha L) | \ \overline{X}^{t_0} > \gamma(C-L), \overline{X}^{t_0} + \overline{X}^{T-t_0} > C - \alpha L]\}$$
$$= \frac{2}{\mu_1 T^2}\{$$
$$\int_0^{\gamma(C-L)} \sigma\{\phi(m(x_1, \beta L)) - m(x_1, \beta L)(1 - \Phi(m(x_1, \beta L)))\}dF_{\overline{X}^{t_0}}(x_1)$$
$$+ \int_{\gamma(C-L)}^{\infty} \sigma\{\phi(m(x_1, \alpha L)) - m(x_1, \alpha L)(1 - \Phi(m(x_1, \alpha L)))\}dF_{\overline{X}^{t_0}}(x_1)\},$$
$$\tag{14}$$

where
$$m(u_1, u_2) = \frac{C - u_1 - u_2 - \mu}{\sigma}, \tag{15}$$

with mean $\mu = \mu_1(T^2 - t_0^2)/2$ and variance $\sigma = \sigma_1^2(T^3 - 3t_0^2 T + 2t_0^3)/3$ of $\overline{X}^{T-t_0}$. ϕ and Φ are the corresponding probability density function and the cumulative probability distribution function for the standard normal $N(0,1)$, respectively.

3. A Seat Inventory Control of Callable Seats with Up-Down Resets

In this section we treat with callable products in the airlines seat inventory model in which the airlines is allowed to cancel the booking seats by paying some compensation to the passengers booked in advance.

3.1. Optimal seat allocation with cancellation

In the model with reset downward in section 2, we assumed that the airline pays no penalty cost for low fare passengers denied for boarding. In this section, we relax this assumption. Let h denote the compensation cost ($h > q$) due to boarding refused. The airlines is allowed to cancel the reservation of low fare passenger due to reset downward at time t_0. The number of the low fare passengers denied for boarding is $(\overline{Y}^{t_0} \wedge L - \alpha L)^+$. So, the total compensation cost is given by

$$h(\overline{Y}^{t_0} \wedge L - \alpha L)^+.$$

Hence, we have the total revenue in reset downward as follows;

$$\overline{v}_2(L, \overline{X}, \overline{Y}) = v_2(L, \overline{X}, \overline{Y}) - (h - q)(\overline{Y}^{t_0} \wedge L - \alpha L)^+. \tag{16}$$

Note that the model can be reduced to the no penalty cost model (equation (6) in section 2.1) when we put $h = q$. The total revenue in the case of reset upward is the same form as equation (5), that is,

$$\overline{v}_1(L, \overline{X}, \overline{Y}) = v_1(L, \overline{X}, \overline{Y}).$$

Defining $\overline{V}(L)$ as the expected total revenue with cancellation over the time interval $[0, T]$, $\overline{V}(L)$ can be given by

$$\overline{V}(L) = V(L) - (h - q)\overline{F}_{\overline{X}^{t_0}}(\gamma(C - L))E[(\overline{Y}^{t_0} \wedge L - \alpha L)^+]. \tag{17}$$

3.2. The model with callable property

In this section, we consider another type of low fare tickets that has a callable property. The callable property is the right for the airlines to possess the option of recalling the seats reserved by the low fare customers at a prespecified recall price. We call those passengers "callable passengers". We assume that there are three types of tickets as follows;

- Type 1 : High fare ticket
  This type ticket is not recalled by the airlines. The fare p is the highest among the tickets.

- Type 2 : Low fare ticket with a callable property
  This ticket has a priority to be recalled for reset downward and the airline pays the recall price d, $q \leq d \leq p$ to the customer where q is the low fare.
- Type 3 : Low fare ticket with no compensation
  For reset downward, this ticket is recalled after the booking of all callable passengers are canceled. And they receive the recall price q. Therefore, there is no compensation since the ticket price is equal to the recall price. If the number of callable passengers are too large then this ticket is not recalled.

Note that the Types 2 and 3 share the same property in the case of reset upward.

To count the number of the callable passengers, we define

$$D_i = \begin{cases} 1 \text{ if the } i\text{th customer makes the decision to grant the call,} \\ 0 \text{ otherwise.} \end{cases}$$

Assume that $\{D_1, D_2, \cdots\}$ are independent and identically distributed with mean $E[D_i] = \delta$ the probability of granting the call to the airlines. If B seats are booked, then $H(B) = \sum_{i=1}^{B} D_i$ seats are confirmed and is binomially distributed with mean $B\delta$. In this case a sequence of operations occurs as follows:

(i) Choose L, α, β, γ, and announce the recall price d.
(ii) Observe the cumulative booking requests $\overline{X}^{t_0}$ from the high fare at the reset time t_0,
   (a) If $\overline{X}^{t_0} \geq \gamma(C - L)$, then L should be reset downward by recalling the tickets. And pay the compensation to the number of passengers given by $\min\{H(\overline{Y}^{t_0} \wedge L), \alpha L\}$ passengers. The net payment to a callable passenger is $d - q \geq 0$.
   (b) If $\overline{X}^{t_0} < \gamma(C - L)$, then L should be reset upward. The option is not exercised.

If $\overline{X}^{t_0} < \gamma(C - L)$, the total revenue $\tilde{v}_1(L, \overline{X}, \overline{Y})$ takes the same form of equation (5);

$$\tilde{v}_1(L, \overline{X}, \overline{Y}) = v_1(L, \overline{X}, \overline{Y}). \tag{18}$$

If $\overline{X}^{t_0} \geq \gamma(C - L)$, the cost of recalling the seats is given by

$$C_L = d \min\{H(\overline{Y}^{t_0} \wedge L), (\overline{Y}^{t_0} \wedge L - \alpha L)^+\}$$
$$+ q((\overline{Y}^{t_0} \wedge L - \alpha L)^+ - H(\overline{Y}^{t_0} \wedge L))^+. \tag{19}$$

The first term is the cost of recalling the Type 2 ticket and the second term the cost of recalling the Type 3 ticket. Hence, the total revenue $\tilde{v}_2(L,\overline{X},\overline{Y})$ is given by

$$\tilde{v}_2(L,\overline{X},\overline{Y}) = v_2(L,\overline{X},\overline{Y}) - (d-q)\min\{H(\overline{Y}^{to} \wedge L), (\overline{Y}^{to} \wedge L - \alpha L)^+\}. \quad (20)$$

When $d = h$ and $\delta = 1$ in equation (20), we get equation (16). The expected total revenue $\tilde{V}(L)$ is given by

$$\tilde{V}(L) = V(L) - (d-q)\overline{F}_{\overline{X}^{to}}(\gamma(C - L))$$
$$\times E[\min\{H(\overline{Y}^{to} \wedge L), (\overline{Y}^{to} \wedge L - \alpha L)^+\}]. \quad (21)$$

In the cancellation model of section 3.1, the airline pays the compensation for all passengers whose tickets are recalled by resetting downward. However, the compensation costs of this section is paid only for passengers called. Therefore, these arguments lead to the following result.

Lemma 4. *If $h \geq d$, the expected revenue with callable ticket $\tilde{V}$ is greater than or equal to the corresponding expected revenue without the callable ticket $\overline{V}$ for any feasible values of L. Furthermore, $\tilde{V}$ can be written as*

$$\tilde{V}(L) = \overline{V}(L) + R(L)$$

where

$$R(L) = \overline{F}_{\overline{X}^{to}}(\gamma(C - L))E[(h - d)(\overline{Y}^{to} \wedge L - \alpha L)^+$$
$$+ (d-q)\{(\overline{Y}^{to} \wedge L - \alpha L)^+ - H(\overline{Y}^{to} \wedge L)\}^+] \geq 0.$$

Proof. From equations (17) and (21), we have

$$\tilde{V}(L) - V(\overline{L}) = E\left[\tilde{v}_2(L,\overline{X},\overline{Y})1_{\{\overline{X}^{to} \geq \gamma(C-L)\}}\right] - E\left[\overline{v}_2(L,\overline{X},\overline{Y})1_{\{\overline{X}^{to} \geq \gamma(C-L)\}}\right]$$
$$= \overline{F}_{\overline{X}^{to}}(\gamma(C - L))E[(h - d)(\overline{Y}^{to} \wedge L - \alpha L)^+$$
$$+ (d-q)\{(\overline{Y}^{to} \wedge L - \alpha L)^+ - H(\overline{Y}^{to} \wedge L)\}^+] \geq 0. \quad \square$$

3.3. A case of low fare demand large enough with callable property

Suppose that there is a sufficiently large amount of low fare demands which are much bigger than the booking limit for low fare demands. We use the normal approximation to the binomial distribution $H(B)$ to make computation simple. Thus, $H(B)$ is normal distribution with mean δB and variance

$\delta(1-\delta)B$. Let $Z(\cdot)$ denote the probability distribution of random variable $H(B)$. In equations (18) and (20) we may delete $\overline{Y}$ from the arguments of function $\tilde{v}_1$ and $\tilde{v}_2$ which can be rewritten as follows;

$$\tilde{v}_1(L,\overline{X}) = v_1(L,\overline{X}) \tag{22}$$

and

$$\tilde{v}_2(L,\overline{X}) = v_2(L,\overline{X}) - (d-q)\min\{H(L), L - \alpha L\}. \tag{23}$$

We have the expected revenue $\hat{V}(L)$ as follows;

$$\hat{V}(L) = V(L) - (d-q)\overline{F}_{\overline{X}^{t_0}}(\gamma(C-L))\left\{(1-\alpha)L - \int_0^{L-\alpha L} Z(z)dz\right\}. \tag{24}$$

4. Numerical Examples

In this section we focus on some numerical results to derive optimal booking limits and to illustrate the revenue function and some impacts of the parameters on the booking policies as well as on the maximum revenues. We then explain economic interpretations of the model implementation.

4.1. Computational results of section 2

Suppose that the high fare demand process is given by Lemma 1. We assume that the airplane has the capacity $C = 300$ and we fix $p = 350$, $q = 100$, $t_0 = 90$, $T = 120$, $\alpha = 0.9$, $\beta = 1.1$ and $\gamma = 0.4$. In Table 1, we compare the revenue function of our model with the one of the classical model ($\alpha = \beta = \gamma = 1$, $t_0 = T$). For example, the expected revenue of our model with the demand parameter $(\mu_1, \sigma_1) = (0.009, 0.05)$ has been improved by 4.78%, compared with the classical model. The optimal booking limit L^* is 226 and the booking policy is as follows ;

The revised booking limit during $[t_0, T] = \begin{cases} \lfloor \beta L^* \rfloor = 248 \text{ seats if } \overline{X}^{t_0} < 29 \\ \lfloor \alpha L^* \rfloor = 203 \text{ seats if } \overline{X}^{t_0} \geq 29. \end{cases}$

The passenger spill rate seems to be in closer agreement with actual airline practice. Also, each spill rate of the reset model is smaller than the one of the classical model. Figure 1 shows the revenue functions for four demand types. The effect of each demand parameter α, β, γ and t_0 is shown by Figures 2–5. Note that the parameters are $C = 300$, $(\mu_1, \sigma_1) = (0.009, 0.05)$, $p = 350$, $q = 100$ and $T = 120$. The percentage terms of each point represents the improvement rate of th expected revenues for the classical model. Moreover, a symbol □ in Figures 2 ($\alpha = 0.7$) and 4

($\gamma = 0.8, 0.9, 1.0$) represents the infeasible solution which does not satisfy the criteria of β. We may observe the fact that L^* is decreasing in β and γ, respectively. Also, L^* is increasing in t_0.

4.2. Computational results of section 3

Let $\mu_1^{t_0}$ and $\sigma_1^{t_0}$ be the expected value and standard deviation of high fare demands in $[0, t_0]$, respectively and μ_1^T and σ_1^T be the expected value and standard deviation of high fare demands in $[t_0, T]$, respectively. We assume that the high fare demands in $[0, t_0]$ is low in terms of the mean, that is, $\mu_1^{t_0} \ll \mu_1^T$. This assumption is more practical since a customer usually purchases the low fare ticket first because of no difference in service between the low fare and high fare tickets.

Furthermore, we assume that a low-fare passenger's decision to seek a booking is based only on his reservation price P_L and the low fare q, not on the recall price d. So, the low fare passengers seek for the booking if and only if $P_L \geq q$. If P_L is uniformly distributed between $[q, p]$, the probability δ that a customers buy the low fare ticket is

$$\delta = \Pr(P_L < d \mid P_L \geq q) = \left(\frac{d-q}{p-q}\right)^+.$$

Suppose that the model parameters are given by $C = 300$, $p = 350$, $q = 100$, $t_0 = 90$, $T = 120$, $\alpha = 0.9$, $\beta = 1.1$, $\gamma = 0.3$ and $d = h = 150$. Using these parameters, we have $\delta = 0.2$. In Table 2, we compare the revenue function of three models (The reset model, callable model and callable model). For example, $(\mu_1^{t_0}, \sigma_1^{t_0}) = (0.008, 0.020)$, $(\mu_1^T, \sigma_1^T) = (0.019, 0.10)$, the expected revenue of callable model has been improved by 0.46%, compared with the cancellation model. On the other hand, L^* of the cancellation model is lower than one of the reset model, by effecting cancellation cost. Figure 6 represents the expected revenue function for four demand types.

5. Conclusion

In this paper, we have formulated the seat allocation model with upward-downward resets for the initial booking limit. Since our model possesses only two periods consisting of a sequential operations with the initial and revised allocations, there are several directions in which future research in this area could be conducted. One would be a dynamic model in which there are n times of the reset opportunity available. Another direction for further work is in developing a model of determining the upward-downward sizes α, β.

Table 1. Comparison of the expected revenue and the spill rates for section 2. (C : classical model, R : reset model, OBL : optimal booking limit, ER : expected revenue, FS : flight spill rate, PS : passenger spill rate).

(μ_1, σ_1)	Case	OBL	ER	%	FS(%)	PS(%)
(0.005,0.03)	C	251	36484.89		28.4	11.2
	R	245	37030.71	1.50	20.2	2.2
(0.009,0.05)	C	214	41923.81		28.8	10.6
	R	226	43928.53	4.78	25.5	4.1
(0.013,0.07)	C	176	47371.64		28.4	10.0
	R	194	50092.38	5.74	27.6	5.7
(0.017,0.09)	C	139	52822.19		28.6	10.0
	R	159	55948.45	5.92	28.6	6.6

Table 2. The impact of revenue performance in section 3 and comparison of models. (CN : with cancellation model, CL : with callable model).

$(\mu_1^{to}, \sigma_1^{to})$	(μ_1^T, σ_1^T)	Case	OBL	ER	%
		R	207	42541.00	
(0.005,0.015)	(0.013,0.07)	CN	205	42399.40	
		CL	205	42436.41	0.09
		R	183	49308.77	
(0.008,0.020)	(0.019,0.10)	CN	176	48997.39	
		CL	182	49221.79	0.46
		R	150	54412.58	
(0.011,0.030)	(0.024,0.17)	CN	144	54074.00	
		CL	150	54516.60	0.82

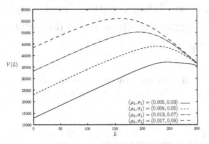

Figure 1. The expected revenue function of booking limit L in section 2.

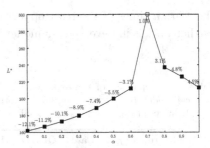

Figure 2. Effect of α
($\beta = 1.1$ and $\gamma = 0.4$).

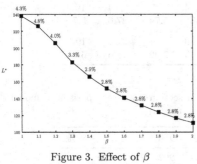

Figure 3. Effect of β
($\alpha = 0.9$ and $\gamma = 0.4$).

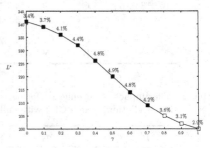

Figure 4. Effect of γ
($\alpha = 0.9$ and $\beta = 1.1$).

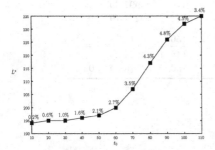

Figure 5. Effect of t_0
($\alpha = 0.9$, $\beta = 1.1$ and $\gamma = 0.4$).

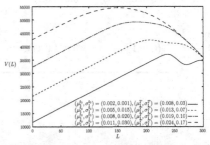

Figure 6. The expected revenue function of booking limit L in section 3.

References

1. K. Sawaki, An Analysis of Airline Seat Allocation, *Journal of Operations Research Society of Japan*, **32**, 411–419, (1989).
2. K. Talluri and G. Van Ryzin, *The Theory and Practice of Revenue Management*, Kluwer Academic Publishers, Boston/Dordrecht/London, (2004).
3. S. L. Brumelle and D. Walczak, Dynamic Airline Revenue Management with Multple Semi-Markov Demand, *Operations Research*, **51**, 137–148, (2003).
4. H. Chao, The EOQ Model with Stochastic Demand and Discounting, *European Journal of Operational Research*, **59**, 434–443, (1992).
5. R. G. Vickson, A Single Product Cycling Problem under Brownian Motion Demand, *Mamagement Sicence*, 32, 1336–1345, (1986).
6. G. Gallego, S. G. Kou and R. Phillips, Revenue Management of Callable Products, *Working paper*, Columbia University, (2004).
7. H. Lee, *Revenue Management of Callable Products with Overbooking*, ProQuest / UMI, (2006).
8. S. L. Brumelle, J. I. McGill, T. H. Oum, K. Sawaki and M. V. Tretheway, Allocation of Airline Seats between Stochastically Dependent Demands, *Transportation Science*, **24** (3), 183–192, (1990).
9. S. M. Ross, Introduction to probability Models, Seventh Edition, (2000).

Part C Reliability and Maintenance

SIMULATION OF RELIABILITY, AVAILABILITY AND MAINTENANCE COSTS

PER-ERIK HAGMARK* AND SEPPO VIRTANEN*

Institute of Machine Design and Operation
Tampere University of Technology
Korkeakoulunkatu 6, FI-33101 Tampere, Finland
per-erik.hagmark@tut.fi, seppo.virtanen@tut.fi

This paper discusses the following six key elements: (1) A quite general stochastic logic generalizes the usual FTA concept. A failure of the TOP-entity (the product in use) is mainly the logical-stochastic short-time consequence of changes in the state of the basic entities. (2) An exceptional rescue mechanism for failed gates enables modelling of simple dynamical features. This improves the practical applicability. (3) A model for deterministic relations between the TOP and the basic entities leads to the existence of two additional intermediate "wait states". (4) The long-time statistical behaviour of each basic entity is modelled with the repair time distribution and a rather general failure point process. (5) A stochastic simulation along the time axis produces a logbook of all events in a chronological order. The effects of age, corrective and preventive maintenance on reliability, availability and maintenance costs are also taken into account. (6) Many examples are given of how the detailed data in the logbook can be used for calculation of figures, graphs, and tables. Along with the conceptual presentation we follow numerical results of a power generation unit which is our case example.

1. Introduction

Since 1996 eleven Finnish companies have participated in the research project conducted by the Tampere University of Technology, whose objective was to develop computer-supported probability based methods for developing reliable and safe systems and equipment. The participating companies are both manufacturers and users of equipment in the metal, energy, process and electronics industries. Their products and systems demand high safety and reliability requirements.

The method presented in this paper is one of the main outcomes from the research project. RAMoptim software has been developed to implement this method. The software is integrated with other software developed in the research project, as indicated in Fig. 1. The failure logic can be designed using ELMAS. The reliability and availability obtained from RAMoptim should be compared with the requirements, set e.g. by using our allocation method RAMalloc. If these or other defined requirements have not been achieved, the designer must return to the drawing table to consider other solutions. Further, our method also enables repair time delays due to external causes. E.g., using StockOptim one can assess the delay caused by the lack of spare parts.

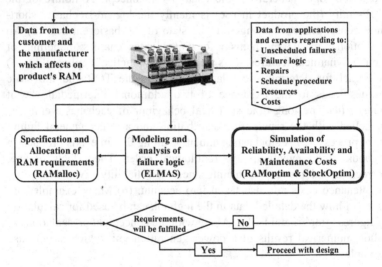

Figure 1. Probabilistic approach to defining reliability and availability requirements for the product and to assess that a proposed design solution fulfills the numeric requirements set for its reliability, availability and maintenance costs.

Ramentor Oy (www.ramentor.com) is responsible for commercializing, marketing and supplying technical support of the related computer software (RAMoptim, StockOptim, ELMAS, RAMalloc, Fig. 1). The software has also been tested at the participating companies. Their experience shows that during the design stage it is possible to identify those problem areas that can delay the product development and/or reduce the safety and reliability.

The application of the methods has guided the companies to transfer their resources from repairing field failures to implementing failure prevention actions during the design stage. Correspondingly, the application of these methods has forced companies to improve their failure reporting and preventive maintenance systems. In addition, the companies have noticed that the engineers need more knowledge in the area of probabilistic approach in reliability and maintainability engineering.

Our method helps the designer to determine at an early stage of the design what level of reliability performance and corrective and preventive maintenance costs can be achieved under the selected design solution, maintenance and operation strategies, and maintenance resources. The versatility of the method also promotes the introduction of expertise from areas that strongly affect the success of the design process, namely the manufacturing, testing, operation and maintenance.

Finally, the development of the method and the related software continues. The effect of preventive maintenance (PM) (e.g. condition monitoring (CM) and diagnostic resources) on failure tendency is a particularly important and a challenging subject for further research.

The outline of the paper is as follows. In Section 2 we define a short-time stochastic logic for failures[1], and a rescue mechanism for gates. Three operation strategies concerning deterministic relationships between the TOP entity (the product) and the basic entities are introduced. This leads to four state alternatives which an entity can have.

In Section 3 we discuss the design for long-time statistical behaviour of the basic entities, i.e., repair time distributions and the point process for failures. Next, the parameters for the effect of age and CM on the occurrence of failures are introduced. Thereafter the simulation process, consisting of simultaneous and interacting "semi-Markov-like" processes (one for each entity), can be started.

The simulation produces a detailed time-ordered logbook of all events, the raw material for subsequent calculation. Section 4 describes some reliability and availability results for the product and the basic entities, and their relations. After additional inputs and the definition of scheduled PM-procedures, Section 5 discusses results on costs and maintenance staff for both CM and PM. Along with the conceptual presentation we follow numerical results of a power generation unit which is our case example. This case is presented in section 2.1.

2. Failure and Operation Structures

2.1. *The case example — power generation unit*

The power generation unit (PGE) consists of two identical generators (GE1 and GE2) and a back-up generator (BUGE). GE1 and GE2 have to run simultaneously in order for PGE to produce required output. In case GE1 or GE2 fails or either of them is taken to the maintenance service, BUGE is called for operation until GE1 or GE2 is repaired or the service is completed. GE1 or GE2 can be repaired or serviced concurrently while the other is still running. In case BUGE fails, the one running is not shut down. The manufacturer and the operators of PGE have specified the following failure and maintenance data for the generators. Costs and resource data related to generators' failure, repair and maintenance service are presented in Table 1.

GE1 and GE2: The generator's mean time to failure (MTTF) is 730 d and deviation 180 d without any preventive maintenance service. Manufacture's recommendation is to take the generator into the specific preventive maintenance service (PMS) after its accumulated running time (H) is 550 d. After PMS is completed, the generator runs as new. Mean PMS time is 2 d and deviation 1 d. Three persons are needed simultaneously to carry out PMS 50% of the time. This means that the average persons needed simultaneously is 1.5. In case the unscheduled failure occurs before H it is repaired and the running time calculation starts from the beginning.

Mean time to repair (MTTR) failed generator is 2.0 d and deviation 1.0 d. Minimum repair time is 0.2 d and maximum 3.5 d, which corresponds to 95% quantile. Probability to have a delay related to start repair (e.g. lack of spare parts) is 80%, mean of delay time is 0.5 d and deviation 0.5 d. According to the repair strategy the generator runs as new after repair. Probability to have failure immediately after repair is 5%. Two persons are needed simultaneously to carry out repair 75% of the time. Naturally, this means that the average persons needed simultaneously is 1.5.

BUGE: The probability that BUGE does not start-up when it is called to operation is 5%. Mean time to repair start-up failure is 0.2 d. One person is needed to repair this start-up failure. Mean time to failure after BUGE is started–up is 360 d. Mean time to repair the failure of BUGE is 1.5 d. According to the repair strategy BUGE runs as new after repair. Two persons are needed simultaneously to carry out the repair of BUGE's running failure 50% of the time.

Table 1. Input costs and resource data.

	PGE	GE1	GE2	PUGE start-up	PUGE running
Loss [€] per failure	10000	0	0	0	0
Down-time costs [€/d]	20000	0	0	0	0
Repair costs per person [€/d]	1200	1200	1200	1200	1200
Repair costs (time independent) [€/failure]	1000	2500	2500	200	1500
PMS costs per person [€/d]	1200	1200	1200	1200	1200
PMS costs (time independent) [€/failure]	1000	2000	2000	200	1500
Number of persons needed for repair	2	2	2	1	2
Share of persons attendance of repair time	0.5	0.75	0.75	1	0.5
Number of persons needed for PMS	0	3	3	0	0
Share of persons attendance of PMS	0	1	1	0	0

2.2. Definition of a stochastic failure logic

The occurrence of failures in a complex product is modeled here as a directed net[1]. A node of the net represents a physical or non-physical *entity*. The *state* of an entity is 0 ('non-failed') or 1 ('failed'). An entity is an *input* of another entity, if its state can have a causal effect on the state of the other entity (Fig. 2, arrows). An entity with input(s) is called a *gate*, an entity with no inputs is a *basic entity* (BE), and if an entity is not an input of any other entity, it is a *TOP-entity*.

The characteristics of a gate is given by the data column (k, m, p, μ, $\pm I_1$, $\pm I_2$, ..., $\pm I_n$) (Fig. 3), where k & m are non-negative integers, $0 \leq p \leq 1$, $\mu \geq 0$, and I_i are the id-numbers of the inputs. The state of a gate is defined as follows. Denote $x_i = state(I_i)$, if $+I_i$, and $x_i = 1-state(I_i)$, if $-I_i$. Then, if $k \leq \Sigma x \leq m$, the state will be 1 with probability p, otherwise 0.

For example, gate ID 9 in Figs. 2 and 3, the data ($k=1$, $m=1$, $p=1$, $+I_1$, $+I_2$, $+I_3$) expresses a XOR (exclusive-OR) gate with two inputs. By varying the parameters k, m, p, and using the NOT operator $(-I_i)$, it is possible to model very complicated stochastic logics [1], for example gate ID 11 in Fig. 2.

In applications, the actual product failure is represented by TOP state 1. Statistically independent events, i.e. basic events, which affect the occurrence of product failure, are interpreted either as BEs in state 1, or (in simple cases) using the gate probability (p), for example, gate ID 10 in Figs. 2 and 3. The failure logic (net) is fully contained in a diagram consisting of one data column for each gate. The graphic of power generation failure logic and the equivalent diagram are presented in Figs. 2 and 3. These are examples for demonstration in subsequent sections. The parameter μ concerns the 'rescue mechanism' to be declared in section 2.4.

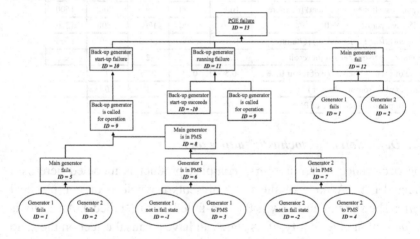

Figure 2. The graphic of power generation failure logic.

ID	5	6	7	8	9	10	11	12	13
k	1	2	2	1	1	1	2	2	1
m	1	2	2	1	1	1	2	2	1
p	1	1	1	1	1	0.05	0.01	1	1
μ	-	-	-	-	-	0.2	1.5	-	-
I_1	1	-1	-2	6	5	9	-10	1	10
I_2	2	3	4	7	8	0	9	2	11
I_3	0	0	0	0	0	0	0	0	12

Figure 3. Failure logic diagram equivalent with the graphic shown in Fig. 2.

2.3. Operation strategies and wait states

Deterministic relations or bounds between TOP and BEs can also exist. Without listing them, we state that many of them can be modeled by setting for each BE to some combination of the following *operation strategies*:

$a=1$: This BE cannot be repaired if TOP is operating. Otherwise $a=0$
$b=1$: This BE is not operating if TOP is not operating. Otherwise $b=0$
$c=1$: TOP will not be started if this BE is still failed. Otherwise $c=0$

Any binary triple (a, b, c) is in principle possible. However, the choice $a=1$ (valid), $c=0$ (negated) can lead to non-practical behavior, where the repair of a BE is interrupted before it is ready. Note that the b-strategy can also be used for gates. The operation strategies of the generators in our example case are the following:

$$\begin{pmatrix} ID & 1 & 2 & 3 & 4 & 5 & 6 & 7 & 8 & 9 & 10 & 11 & 12 \\ a & 0 & 0 & 0 & 0 & & & & & & & & \\ b & 0 & 0 & 0 & 0 & 0 & 0 & 0 & 0 & 0 & 0 & 0 & 0 \\ c & 0 & 0 & 0 & 0 & & & & & & & & \end{pmatrix} \quad (1)$$

So far, we have defined the states 0 & 1 for entities. The operation strategies imply the existence of two new states, the *wait states* (0.5 & 1.5). Thus, we have the complete set of states for entities:

0   Non-failed and operating
0.5 Non-failed but start is denied
1   Failed and available for repair
1.5 Failed but repair is denied (only for BEs)

For example, when a repair of a BE is finished, the state changes from 1 to 0 or 0.5, and when the BE fails, its state changes from 0 to 1 or 1.5. In general, a change in the state of an entity is always caused by the appearance or disappearance of failure somewhere in the net.

Note that scheduled or pre-planned stops (of any kind) are not wait states. These can, as well as the corresponding unavailability, be taken into account beforehand or perhaps afterwards.

2.4. Rescue of a gate

We introduce still another feature that extends practical applicability. The parameter $\mu \geq 0$ in the data column of a gate (Sec. 2.1, Fig. 3) defines the *rescue property*. If $\mu = 0$, the gate always obeys the failure logic, but if $\mu > 0$, the gate possesses the possibility to be rescued after some time from state 1 to state 0 - *against* the failure logic.

Suppose one or more gates with $\mu > 0$ have just entered state 1, i.e. they have failed according to the failure logic (Sec. 2.1). For each of these gates, a random rescue time with mean μ is now generated. When the shortest of these times ends, the gate in question goes into state 0, *unless* a change in the state of some BE has taken place before that. Irrespective of which alternative occurred first, all ongoing rescue efforts stop or are ignored.

Further, if the quickest rescue took place first, then only the states of logically *higher* gates need to be computed (according to the failure logic), but if some BE changed first, *all* gates are recomputed. Also note that in both cases TOP may change state, and this in turn can, according to operation strategies (Sec. 2.2), affect BEs' wait states, usage amounts, and thereby the subsequent behavior.

In the failure logic diagram of the power unit (Fig. 3), parameter μ is specified for BUGE's start-up failure (ID 10) and running failure (ID 11) (Sec. 2.1). In case BUGE does not start-up, its mean time to repair $\mu = 0.2$ d, and after repair is completed BUGE is started up again. Respectively, mean time to repair BUGE's running failure $\mu = 1.5$ d. Parameter $p = 0.05$ for BUGE's start-up failure is specified in Sec. 2.1. In the failure logic diagram (Fig. 3) parameter $p = 0.01$ for ID 11 illustrates the probability that BUGE fails during the repair or while preventive maintenance is been performed for GE1 or GE2. This p is calculated based on the BUGE's MTTF (= 360 d) and GE1's maximum repair time (95% quantile = 3.5 d).

3. Dependability Data and Simulation

3.1. *Repair time*

A BE's repair time is by definition the length of a single period in state 1. The related software offers several methods to build the required cumulative distribution function (Cdf). Modifications of standard distributions, mixtures, splines and other tailor-made constructions have been developed to guarantee sufficient flexibility. Depending on the sort of available input data, various combinations of means, deviations, medians, quantiles, minimums, maximums, censoring, weights, experts' competence, etc. can be used.

Repair time can include delays from external causes. The software also supports separate adding of delays. Note that delays caused by operation strategies (state 1.5) are "internal" and not of this kind. The lack of repair staff can be of this kind if it can be assessed beforehand. Further, the lack of spare parts is of this kind, and the corresponding delay can be assessed with another module of the software (StockOptim)[2]. From input data (Sec. 2.1) modeled cumulative distribution function of GE1's and GE2's total repair time is shown in Fig. 4.

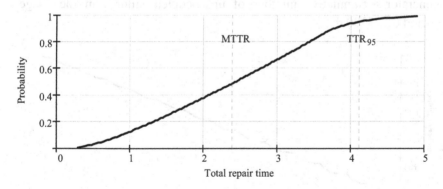

Figure 4. Probability function of total repair time of GE1 and GE2.

3.2. *Failure tendency and failure profile*

Each BE has usually a natural own 'unit of usage' (uu) for the measuring of operative usage. The *usage profile* is a non-decreasing function, whose value $x = \Psi(t)$ is the average amount of usage achieved by a BE in the age period $0...t$ (tu), tu = time unit. Note that t only cumulates the BE's own operative time (state = 0), so $\Psi(t)$ is not directly a plan for usage during the design period. Before the simulation (Sec. 3.4), it is not even known how the operative time of the BE grows during the design period.

The *failure tendency* of a BE is defined with a non-decreasing function, whose value $v = \Lambda(x)$ expresses the cumulative average number v of failures in the usage period $0...x$ (uu). Depending on the nature of available data, our software provides several methods for the design of $\Lambda(x)$. Supported types of data are the number of failures, failure rates, reliability, availability, censoring, life distributions, etc. One can also experiment with changes in the length of service period for exchangeable BEs.

Finally, the combination of the usage profile and the failure tendency yields the *failure profile* $v = \Lambda(\Psi(t))$, i.e. the average number of failures occurring in the age period $0...t$ (tu). In our example (Sec. 2.1), generators GE1's and GE2's cumulative average number of failures in the usage period $0...7300$ d is shown in Fig. 5. This failure tendency curve illustrates the effect of the specific PMS (Sec. 2.1) on the generator's cumulative number of unscheduled failures in the usage period $0....7300$ d.

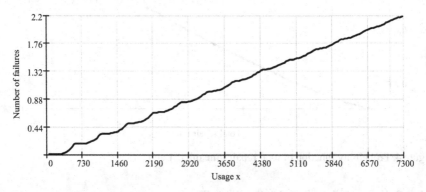

Figure 5. Modeled failure tendency function from input data for GE1&2.

3.3. Generation of failure points

Caused failure and other stops of BEs can affect the subsequent behavior. It is assumed that these effects can be interpreted as 'changes' in the age t of the BE or in the probability to fail immediately after repair. The model offers the following parameters:

$L \geq 0$ Age of the BE at original start
$0 \leq P \leq 1$ Failure probability immediately after repair
$Z \geq 0$ Change of age caused by stopping (state $0 \rightarrow 0.5$); NewAge = $Z \cdot$OldAge
$Y \geq 0$ Change of age caused by repair; NewAge = $Y \cdot$OldAge

The main input to the random generation of failure points (x) is a non-decreasing function $g(x)$. This together with different choices of L, P, Z and Y covers a rich variety of point processes. Without details we mention two important special cases here: First, supposedly the BE is 'new' at the start ($L = 0$), no immediate failure occurs after repair ($P = 0$), and there is no effect from stopping without failure ($Z = 1$). Then, (a) for $Y = 1$ (minimal repair) and $g(x) := \Lambda(x)$ (Sec. 3.2) our generation method produces a non-homogeneous Poisson process, and (b) for $Y = 0$ (maximal repair) and $g(x) := -\ln(1-F(x))$, where F is the cumulative distribution function of the first failure, we get a renewal process (where $\Lambda \neq g$, in general).

3.4. Simulation procedure and the logbook

In the beginning all BEs are working (state 0), and a random time to failure is generated for each BE (Sec. 3.2-3). When the first BE fails, the states of the gates are generated up to TOP according to Sec. 2. If TOP is still working (state 0) and $a=1$, the failed BE takes the "waiting-for-repair" state 1.5; otherwise (TOP failed or $a=0$) the failed BE takes the state 1, and repair time is generated (Sec. 3.1). Besides, if TOP failed (state 1), then those working (non-failed) BEs, whose $b=1$, must go into the "waiting-for-start" state (0.5), etc.

Mathematically speaking, the simulation consists of interacting 4-state (Sec. 2.3) "semi-Markov-like" processes, one for each entity. This goes on until the end of the design period, i.e., until the operating time of TOP reaches a predefined value, T (tu). All this is documented in a 'logbook' describing events, time instants, TOP-age instants, states, duration of state combinations, etc.

A direct and detailed study of the logbook is the designer's best preliminary assessment for how successfully the model and the input data match the desired behavior. The logbook in Table 2 is the raw data of our case example. The first row shows the number of simulations ($N = 2000$), design life, i.e., TOP-interval ($T = 7300$), TOP ($ID = 13$), other gates ($ID = 5, 6, 7, 8, 9, 10, 11, 12$) and BEs ($ID = 1, 2, 3, 4$). The following rows describe, in chronological order, the behavior of the built model: cumulative total time, TOP-age, state combination for the entities 13, 1, 2, 3, 4, 5, 6, 7, 8, 9, 10, 11, 12 and time duration of this combination.

Table 2. The first page of the logbook (Figs. 2 and 3) (Description in the main text)

2000	7300	13	1	2	3	4	5	6	7	8	9	10	11	12	0
0	0	0	0	0	0	0	0	0	0	0	0	0	0	0	374.60
374.60	374.60	0	0	0	0	1	0	0	1	1	1	0	0	0	2.90
377.50	377.50	0	0	0	0	0	0	0	0	0	0	0	0	0	470.40
847.90	847.90	0	0	0	1	0	0	1	0	1	1	0	0	0	3.90
851.70	851.70	0	0	0	0	0	0	0	0	0	0	0	0	0	364.30
1216.00	1216.00	0	0	0	0	1	0	0	1	1	1	0	0	0	4.50
5766.20	5765.30	0	0	0	0	0	0	0	0	0	0	0	0	0	347.40
6113.50	6112.60	1	0	0	1	0	0	1	0	1	1	0	1	0	0.60
6114.10	6112.60	0	0	0	1	0	0	1	0	1	1	0	0	0	0.70
6114.80	6113.30	0	0	0	0	0	0	0	0	0	0	0	0	0	427.70
6542.40	6541.00	0	0	0	0	1	0	0	1	1	1	0	0	0	1.20
6543.60	6542.20	0	0	0	0	0	0	0	0	0	0	0	0	0	265.80

The first extract from the logbook tells how much time the entities spend in state 0 during the TOP-interval $0...T$. Exactly, for each entity, this is the sum of those time intervals in the right column during which the entity is in state 0 divided by number of simulations. The numerical result for the case example is:

$$\begin{pmatrix} 13 & 1 & 2 & 3 & 4 & 5 & 6 & 7 & 8 & 9 & 10 & 11 & 12 \\ 7300 & 7294.8 & 7294.9 & 7279 & 7278.4 & 7289.2 & 7279 & 7278.4 & 7256.6 & 7245 & 7300.3 & 7300.3 & 7301 \end{pmatrix}$$

(2)

4. Reliability and Availability Calculation

4.1. *Main results for the product*

Various measures and indicators for reliability and availability can be extracted from the logbook. We consider three important concepts for TOP. A *downtime* period is an unbroken non-operative period (state $\neq 0$). Thus, a TOP downtime period consists of states 0.5 or 1, since TOP never enters the state 1.5 (Sec. 2.2). Since downtime is always caused by failures, the *failure profile* of TOP is interpreted as the cumulative number of downtime periods during the design period $0...T$ (cp. Sec. 3.2). Finally, the *point wise availability* of TOP is a combination of downtime and failure profile. The graphs from downtime, failure profile and availability of the power generator unit are shown in Figs. 6, 7 and 8.

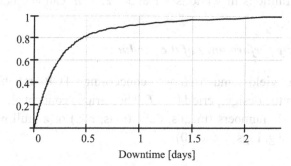

Figure 6. Cdf for TOP single downtime; mean 0.334, de viation 0.481.

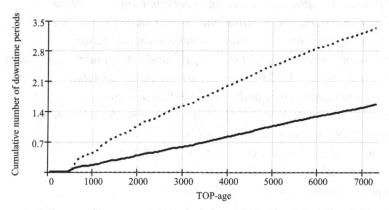

Figure 7. Failure profile; mean 1.582, deviation 1.244. The dotted line is the 95% quantile profiles.

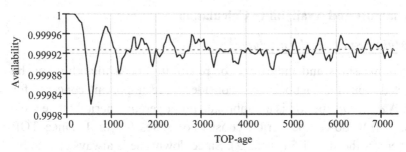

Figure 8. Point wise availability; mean 0.999928.

These results can be compared with possibly existing reliability and availability requirements, which have perhaps been assessed by using some allocation method, e.g. Ref. 3. Moreover, since these functions also match the definitions in sections 3.1 and 3.2, TOP can act as a BE in a higher net.

4.2. Reliability performance of the product

Our method yields much more concerning TOP reliability and availability in the design period $0 \ldots T$. The various concepts listed below appear as single numbers (means, deviations, etc.) or as full probability distributions (e.g. Figs. 9 and 10).

Duration and number of downtime and corresponding deviations
Duration and number of repairs and corresponding deviations
Duration and number of waiting and corresponding deviations
Duration of downtime period and corresponding deviation
Duration of operation period (MTTF) (0) and corresponding deviation
Duration of repair (MTTR) (1) and corresponding deviation
Duration of waiting (0.5) and corresponding deviation
Mean time to first failure (MTTFF) and corresponding deviation
Maximum (95%) duration of repair (1)
Maximum (95%) duration of downtime period (0.5 or 1)
Failure probability within desired interval $t_1...t_2$
Average unavailability (caused by failures)

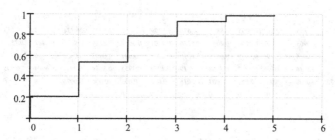

Figure 9. Cdf of the number of downtime periods; mean 1.582, deviation 1.244.

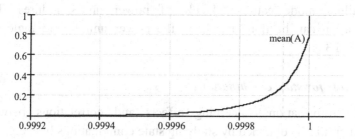

Figure 10. Total availability; mean 0.999928, deviation 0.000103.

4.3. Entities involved in TOP downtime

An entity's role in TOP downtime (Sec. 4.1) can be described from various points of view. Our method provides the following five indicators.

- N   the number of failures of this entity
- P1 probability that a failure of this entity (1 or 1.5) starts a TOP downtime period
- P2 probability that this entity is already failed when a TOP downtime period starts
- P3 time proportion of entity's failures in TOP downtime
- U1 unavailability due to states 1 & 1.5
- U2 unavailability due to state 0.5

The numerical result of entities' role in the power generator unit's failure is:

$$\begin{pmatrix} ID & 13 & 1 & 2 & 3 & 4 & 5 & 6 & 7 & 8 & 9 & 10 & 11 & 12 \\ N & 1.582 & 2.368 & 2.31 & 10.8005 & 10.8825 & 4.678 & 10.799 & 10.8835 & 21.6825 & 26.334 & 1.3245 & 0.253 & 0.004 \\ P1 & 1 & 0.0963 & 0.0839 & 0.4104 & 0.4047 & 0.1776 & 0.4101 & 0.4047 & 0.8176 & 0.9952 & 0.8441 & 0.1534 & 0.0026 \\ P2 & 0 & 0.0026 & 0.0019 & 0.0029 & 0.0013 & 0.0019 & 0.0022 & 0.001 & 0.0003 & 0.0022 & 0 & 0 & 0 \\ P3 & 1 & 0.1136 & 0.0895 & 0.4087 & 0.3958 & 0.19 & 0.4079 & 0.3955 & 0.8034 & 0.9934 & 0.4904 & 0.503 & 0.0066 \\ U1 & 0.000072 & 0.0008 & 0.0008 & 0.003 & 0.003 & 0.0016 & 0.003 & 0.003 & 0.006 & 0.0076 & 0 & 0 & 0 \\ U2 & 0 & 0 & 0 & 0 & 0 & 0 & 0 & 0 & 0 & 0 & 0 & 0 & 0 \end{pmatrix}$$

(3)

From the results (3) row *P1* we can see that the most probable cause lead to power failure (ID 13) which is BUGE start-up failure (ID 10). This failure mode causes 84.41% of power unit's failures. This interpretation can be done based on the power unit failure logic, see Figs. 2 and 3.

4.4. Results for state combinations

Valuable information concerning effects and correlations between different entities is obtained by studying state combinations like, e.g.,

$$\begin{pmatrix} ID & ID & .... \\ State & State & .... \end{pmatrix} = \begin{pmatrix} 3 & 10 & .... \\ 0 & 1.5 & .... \end{pmatrix}.$$

If T' denotes the total time of the system (first column of Table 2) and T is the operating time of TOP (second column of Table 2), the following figures are given for any selected state combination:

Time   combinations total time during design period $0...T$
Prop   time proportion ($= Time / T'$)
Occ   the number of occurrences during design period

These results can reveal e.g. undesirably long wait states (0.5 or 1.5), or statistical correlation if some combinations occur very often or lasts very long. In our case example interesting is to find out how often and for how long time back-up generator (BUGE, ID 11) is running during the power generation usage period $0...7300$ d. Back-up generator (BUGE) is called to operation always when either GE1 or GE2 has failed, or taken into the maintenance service (PMS) (ID-numbers 1, 2, 3, 4). A numerical result is shown in Table 3. From lower part of Table 3, it can be seen that almost 82% of power unit failures occur while PMS is executed for GE1 or GE2 (ID 3, ID4).

Table 3. Number and duration of BUGE running periods

	ID	1	2	3	4	11	13	Occurrence	Duration
Combination		1	0	0	0	0	0	2.36	5.67
		0	1	0	0	0	0	2.31	5.51
		0	0	1	0	0	0	10.77	21.63
		0	0	0	1	0	0	10.84	21.83
							Total	26.27	54.64
	ID	1	2	3	4		13	Occurrence	Duration
Combination		1	0	0	0		1	0.15	0.056
		0	1	0	0		1	0.13	0.043
		1	1	0	0		1	0.004	0.003
		0	0	1	0		1	0.65	0.214
		0	0	0	1		1	0.64	0.207
							Total	1.58	0.52

5. Cost and Resource Calculation

5.1. *Additional inputs concerning failures*

The logbook (Table 1) constitutes a list of all events caused by failures, and thereby it is the raw material for the calculation of failure caused costs and resources. The designer is next asked for the following additional input data for each entity:

    Lf loss caused by (one) downtime (€)
    Ln loss caused by downtime (€/tu)
    D downtime (tu) without costs
    R number of persons needed for repair (average)
    G time independent repair cost (€)
    H cost per repair time unit (€/tu)

Note that this data is also for gates (including TOP), so that any additional loss due to the gate failure itself can be taken into account.

5.2. *Costs and resources due to failures*

Based on the cost and loss input of Table 1, our model handles the following ten concepts and their numerical values.

0   ID-number for entity
1   Operating time, 0 (tu)
2   Waiting for start, 0.5 (tu)
3   Failed and available for repair, 1 (tu)
4   Failed but waiting for repair, 1.5 (tu)
5   Duration of downtime, 0.5, 1, 1.5 (tu)
6   Number of failures
7   Time independent repair cost (€)
8   Time dependent repair cost (€/tu)
9   Loss caused by failure (€)
10 Loss caused by downtime (€/tu)

The numerical results for the power generation unit are given in (4). Note that in this case preventive maintenance, PMS for GE1 and GE2 (ID 3 and 4), is handled as a corrective maintenance because it is a remarkable issue to cause unscheduled failure for the operation of the power generation unit. In practice, rows 5 and 6 in the numerical results (4) include number and duration of PMS for GE1 and GE2 (ID 3 and 4); correspondingly rows 7 and 8 include all costs of PMS for GE1 and GE2.

$$\begin{pmatrix} 0 & 13 & 1 & 2 & 3 & 4 & 5 & 6 & 7 & 8 & 9 & 10 & 11 & 12 & \text{Sum} \\ 1 & 7300 & 7295 & 7295 & 0 & 0 & 0 & 0 & 0 & 0 & 0 & 0 & 54.6 & 0 & \\ 2 & 0 & 0 & 0 & 0 & 0 & 0 & 0 & 0 & 0 & 0 & 0 & 0 & 0 & \\ 3 & 0.52 & 5.76 & 5.59 & 21.95 & 22.14 & 0 & 0 & 0 & 0 & 0 & 0 & 0 & 0 & \\ 4 & 0 & 0 & 0 & 0 & 0 & 0 & 0 & 0 & 0 & 0 & 0 & 0 & 0 & \\ 5 & 0.52 & 5.76 & 5.59 & 21.95 & 22.14 & 11.35 & 21.93 & 22.13 & 43.91 & 55.19 & 0.26 & 0.26 & 0.003 & \\ 6 & 1.58 & 2.37 & 2.31 & 10.80 & 10.88 & 4.68 & 10.80 & 10.88 & 21.68 & 26.33 & 1.32 & 0.25 & 0.004 & \\ 7 & 1582 & 5920 & 5775 & 21601 & 21765 & 0 & 0 & 0 & 0 & 0 & 265 & 380 & 0 & 57287 \\ 8 & 629 & 10372 & 10068 & 39506 & 39860 & 0 & 0 & 0 & 0 & 0 & 308 & 316 & 0 & 101059 \\ 9 & 15820 & 0 & 0 & 0 & 0 & 0 & 0 & 0 & 0 & 0 & 0 & 0 & 0 & 15820 \\ 10 & 10478 & 0 & 0 & 0 & 0 & 0 & 0 & 0 & 0 & 0 & 0 & 0 & 0 & 10478 \end{pmatrix}$$

(4)

Next we would like to know how much staff is needed for corrective maintenance. We consider the following concepts and the numeric result for our example case (5). As previous results (4), this resource calculation also includes resources related to preventive maintenance, PMS for GE1 and GE2.

Pr time with Rn persons
Rn number of persons

Rr average number of persons needed simultaneously
PWR total person-work time

$$\begin{bmatrix} Pr & 8.4 & 21.7 & 0.1 & 0.2 & 0.1 & PWR & Rr \\ Rn & 2 & 3 & 5 & 6 & 7 & 84.22 & 2.76 \end{bmatrix} \quad (5)$$

From (5), it can be seen how much of work time is totally spent on repair (*PWR* = 84.22 d), and how much staff is needed on average (*Rr* = 2.8). The total time for exactly 3 persons is 21.7 d. A practical conclusion could perhaps be that a staff of 3 persons would result in acceptable delays.

Finally, note that the concepts in this section match the inputs in section 5.1, so TOP can be a BE in a higher net (cp. Sec. 4.1).

5.3. Documentation of preventive maintenance

The connection between failure tendency and preventive maintenance (PM) is handled here on a very heuristic (and optimistic) level: We assume that the designer is able to assess the scheduled procedures needed to maintain the failure tendency designed (Sec. 3.2).

Our model allows the designer to define different scheduled procedures, *SP* = 1, 2..., for each BE. For example, if the BE is a car, then *SP* = 1 can denote "oil changes", *SP* = 2 can be "the large service package for a fairly new car", and *SP* = 3 "the large service package for an old car", etc. The SPs performed at the same time form an *SP-group*.

The *schedule* for SP-groups is then built up, row by row. Table 4 is the schedule for our example. Each row carries the information of an SP-group. The TOP-age is the cumulative operating time of TOP. The STOP column shows the sign 0 or 1 according to whether TOP can or cannot be operating during this SP-group. The next five columns are:

- R persons needed simultaneously (average)
- μ average duration (tu)
- σ deviation of duration (tu)
- G time independent costs (€)
- H time dependent costs (€/tu)

The format of the columns *BE.SP* means that the procedure *SP* is performed to the basic entity *BE*. For example, 1.1 means that *BE* = 1 will be addressed by its *SP* = 1.

The growth of a BEs' own usage is of course essential when planning the service schedule. To this purpose, one must first recall the usage and failure profile (Sec. 3.2). In our example, simulation indicates that about 11 PMS for GE1 and GE2 will be carried out during the power generation usage period 0....7300 d. The simulation based on the manufacturer's failure data and recommendation of PMS interval, see Sec. 2.1. This means that real mean time to PMS is about 650 days which is shown in TOP-age in Table 4.

Table 4. A schedule for constant procedures.

TOP-age	Stop	R	μ	σ	G	H	BE.SP
500	0	3	2	1	2000	1200	3.1
700	0	3	2	1	2000	1200	4.1
1150	0	3	2	1	2000	1200	3.1
1350	0	3	2	1	2000	1200	4.1
1800	0	3	2	1	2000	1200	3.1
2000	0	3	2	1	2000	1200	4.1
2450	0	3	2	1	2000	1200	3.1
2650	0	3	2	1	2000	1200	4.1
3100	0	3	2	1	2000	1200	3.1

5.4. Costs and resources due to scheduled procedures

The schedule is the frame for cost and resource accounting. Our model offers e.g. the following concepts and their numerical values.

GG Time independent costs (€)
HH Time dependent costs (€)
DurS Total duration of SPs (tu)
NonS Non-operating time
AvS Average unavailability
PrS Time with RnS persons
Rs Average number of persons needed simultaneously
PWS Total person-work time

$$\begin{pmatrix} GG & HH & DurS & NonS & AvS & PWS & Rs \\ 43366 & 79366 & 44.09 & 0 & 0.000058 & 66.14 & 1.5 \end{pmatrix} \quad (6)$$

In our case example, costs of PMS (6) are also included in results (4). Total unavailability is 0.000072 (3) from which PMS causes 0.000058.

6. The Summary of the Case Example

Finally, in Table 5 is a short summary of the numeric results for our case example sec. 2.1. The information is combined from (4), (5) and (6).

Table 5. Summary of availability, reliability, costs and resources.

Power generation unit running time 7300 d		
Total availability	0.999928	
Unavailability caused by failures		0.000014
Unavailability caused by SPs		0.000058
Reliability in age period $t_1 ... t_2$ (e.g. 0....730)	0.8760	
Number of failures & deviation	1.582 & 1.244	
Duration of repair & deviation	0.52 & 0.75	
Mean time to first failure & deviation	3803.8 & 2456.7	
Mean time to failure (0....7300) & deviation	2842.7 & 2197.8	
Total maintenance costs	184644	
Costs caused by failures	61912	
Loss caused by failure		15820
Loss caused by downtime		10478
Time independent repair costs		13921
Time dependent repair costs		21693
Costs caused by SPs	122732	
Time independent SP-costs	43366	
Time dependent SP-costs	79366	
Required maintenance resources		
Total work time	84.22	
Caused by failures	18.08	
Caused by SPs	66.14	
Persons needed simultaneously	2.76	

References

1. S. Virtanen, P-E. Hagmark, J-P Penttinen, Modelling and Analysis of Causes and Consequences of Failures. *Annual Reliability and Maintainability Symposium* (RAMS). January 23–26, 2006. Newport Beach, CA, USA.
2. P-E. Hagmark, H. Pernu, Risk Evaluation of a Spare Part Stock by Stochastic Simulation. *ESREL 2006 Conference, Safety and Reliability for Managing Risk.* 18–22 September 2006 – Estoril, Portugal.
3. S. Virtanen, P-E. Hagmark, Specification and Allocation of Reliability and Availability Requirements. *Annual Reliability and Maintainability Symposium* (RAMS). January 23–26, 2006. Newport Beach, CA, USA.

STOCHASTIC PROFIT MODELS UNDER REPAIR-LIMIT REPLACEMENT PROGRAM

TADASHI DOHI[†], NAOTO KAIO[‡] AND SHUNJI OSAKI[††]

[†] *Department of Information Engineering*
Graduate School of Engineering, Hiroshima University
1-4-1 Kagamiyama, Higashi-Hiroshima 739-8527, Japan
dohi@rel.hiroshima-u.ac.jp
[‡] *Department of Economic Informatics*
Faculty of Economic Sciences, Hiroshima Shudo University
1-1-1 Ozukahigashi, Asaminami-Ku, Hiroshima 731-3195, Japan
[††] *Department of Information and Telecommunication Engineering*
Faculty of Mathematical Sciences and Information Engineering
Nanzan University, 27 Seirei-Cho, Seto 489-0863, Japan

In this paper, we consider two repair-time limit replacement problems under earning rate criteria with and without discounting. First, the optimal repair-time limits, which maximize the long-run average profit rate and the expected total discounted profit over an infinite time horizon, are analytically derived. Next, we develop statistically non-parametric algorithms for estimating the optimal repair-time limits, provided that the complete sample data of repair time are given. The basic idea is to apply the Lorenz statistics and to transform the underlying algebraic optimization problems to the graphical ones.

1. Introduction

Since the seminal contribution by Hastings [10], a huge number of repair limit replacement problems have been considered in the literature (*e.g.*, see [6], [12], [13] and [14]). Similar to Nakagawa [13], consider a simple repair-time limit replacement problem under earning rate criteria. Suppose that a production machine with a part is operating, where if the part fails, the repair or replacement is planned. More specifically, when the part fails, the decison maker (DM) estimates the completion time distribution of repair, which may be a possibly subjective one. If DM judges that the repair may be completed up to a prespecified time-limit, the repair operation starts immediately just after the occurrence of failure, otherwise, the spare part is ordered and the failed part is replaced by the new one. Then the problem for DM is to determine the optimal repair-time limit which maximizes an earning rate criterion.

Dohi et al. [2] considered a stochastic profit model under an earning rate criterion to the above one. However, the repair-limit replacement program in [2] was rather different from ours, because their repair-limit replacement policy was based on the stopping rule to give up repairing the failed part. On the other hand, Dohi et al. [3], [7] introduced the concept of subjective repair-time distribution, and discussed the expected cost minimization problems to estimate the optimal decision on repair or replacement. Recently, the same authors [4] formulated the similar but somewhat different repair-time limit replacement problem from the above. The main difference in [4] from this paper is that the imperfect repair is assumed. In other words, the Refs. [3] and [4] considered the same repair-limit replacement problems with different criteria and imperfect repair. In that sense, this paper may be regarded as a continuation of Ref. [7]. For a good survey of repair-limit replacement problems, see Dohi and Kaio [6].

In a fashion similar to the previous papers [2],[3],[4],[7], we concern the statistical estimation problem on the optimal repair-time limit. In [2], it was shown that the optimal repair-time limit can be estimated by means of the usual total time on test statistics. Since the resulting estimator is a non-parametric one with strong consistency, it would be possible to estimate the optimal repairtime limit without specifying the repair-time distribution [5]. While, in Refs. [3],[4],[7], the authors introduced the Lorenz statistics and derived a novel estimator of the repair-time limit. The Lorenz curve was first introduced by Lorenz [11] to describe income distributions. Since the Lorenz curve is essentially equivalent to the Pareto curve used in the quality control, it has been considered as one of the most important statistics applied in sciences, engineering and management. The more general and tractable definition of the Lorenz curve was made by Gastwirth [8]. Goldie [9] proved the strong consistency of the empirical Lorenz curve and discovered its several convergence properties. Chandra and Singpurwalla [1] investigated the relationship between the total time on test statistics and the Lorenz statistics, and derived a few aging and partial ordering properties.

The rest part of this paper is organized as follows. After describing the notation and assumptions used here, we explain the repair-time limit replacement problem under consideration in Section 2. In Section 3, we formulate the long-run average profit rate and the expected total discounted profit over an infinite time horizon, and derive the optimal repair-time limits which maximize them analytically. Next, the algebraic optimization problems are transformed to the geometrical ones in Section 4, where the dual theorems on the optimal repair-time limits are given. In Section 5, we develop the statistically non-parametric algorithms for estimating the optimal repair-time limits, provided that the complete sample data of repair time are given. Since the knowledge on the repair-time distribution is incomplete in general, such a statistical estimation method for the optimal repair-time limit will be useful in the practical maintenance management.

2. Repair-Time Limit Replacement Model

2.1. *Notation*

The repair time Y of each failed part is an independent and identically distributed non-negative random variable. The decision maker (DM) has a *subjective* cumulative distribution function (c.d.f.) $\Pr\{Y \leq t\} = G(t)$ on the repair completion time, with probability density function (p.d.f.) $g(t)(> 0)$ and finite mean $1/\lambda(> 0)$. Suppose that the c.d.f. $G(t) \in [0, 1]$ is arbitrary, absolutely continuous and strictly increasing in $t \in (0, \infty)$. In addition, it is assumed that $G(t)$ has an inverse function, *i.e.*, $G^{-1}(\cdot)$. Further, we define:

- $t_0 \in [0, \infty)$: repair-time limit (decision variable)
- $F(t)$, $f(t)$, $1/\mu$ (> 0): c.d.f., p.d.f. and mean of time to failure of a part
- k (> 0): penalty cost per unit time when the production machine is in down state
- e_0 (> 0): earning rate per unit operation time
- e_1 (> 0): repair cost per unit time
- c (> 0): fixed cost associated with the ordering of a new part
- L (> 0): lead time for delivery of a new part
- β (> 0): discount rate
- $\mathcal{L}\{\psi(\beta)\} := \int_0^\infty \exp(-\beta t)\psi(t)dt$ for an arbitrary continuous function $\psi(\cdot)$ (Laplace transform of $\psi(\cdot)$)
- $\overline{\psi}(\cdot) := 1 - \psi(\cdot)$ (survivor function).

2.2. *Model description*

Consider a simple production machine with a part, where each failed part is repairable but may be provided after a lead time L (> 0) if it is replaced by a new one. The original production machine begins operating at time

$t = 0$. The mean lifetime X of each part is the non-negative random variable having the mean time to failure (MTTF) $1/\lambda$ (> 0). Once the machine failed, the decision maker (DM) wishes to know whether he or she should repair it or order a new spare. If DM estimates that the repair is completed within a prespecified time limit $t_0 \in [0, \infty)$, then the repair starts immediately at the failure time $t = X$ and completes it at $t = X + Y$, where the repair time Y of each failed part has an arbitrary continuous probability distribution $G(t) = \Pr\{Y \leq t\}$ with mean time to repair (MTTR) $1/\mu$ (> 0).

Without any loss of generality, it is assumed that $G(0) = 0$ and $\lim_{t \to \infty} G(t) = 1$. After the completion of repair operation, the repaired part can become as good as new and the production machine equipped with the part starts operating again. On the other hand, if DM estimates that the repair time exceeds the time limit t_0, then the failed part is scrapped at time $t = X$ and a new spare part is ordered immediately. After the new part is delivered after the lead time L, the failed part is replaced by the new one and the production machine starts operating at time $t = X + L$ again, where the time required for replacement of a failed unit can be negligible. Under the above model setting, we define the time interval from the star of operation to the next starting point as one cycle. Suppose that the same cycle is repeated again and again over an infinite time horizon. Figure 1 depicts the configuration of the repair-time limit replacement problem under consideration.

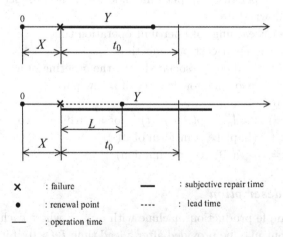

Figure 1. Configuration of repair-time limit replacement with imperfect repair.

3. Analysis

3.1. Long-run average profit

Since the starting point of machine operation can be regarded as a renewal point, the mean time length of one cycle is given by

$$T_L(t_0) = 1/\mu + \int_0^{t_0} t\,dG(t) + \int_{t_0}^{\infty} L\,dG(t). \tag{1}$$

Also, the expected total profit during one cycle becomes

$$V_L(t_0) = e_0/\mu - (e_1 + k)\int_0^{t_0} t\,dG(t) - k\int_{t_0}^{\infty} L\,dG(t) - c\int_{t_0}^{\infty} L\,dG(t). \tag{2}$$

From the familiar renewal reward argument, the long-run average profit rate (the expected total profit per unit time in the steady state) is given by

$$TP_L(t_0) = \lim_{t\to\infty} \frac{\mathrm{E}[\text{total profit on } (0,t]]}{t} = \frac{V_L(t_0)}{T_L(t_0)}, \tag{3}$$

and the problem is to derive the optimal repair-time limit $t_0^* \in [0,\infty)$ satisfying

$$TP_L(t_0^*) = \max_{0 \le t_0 < \infty} TP_L(t_0). \tag{4}$$

We make the following assumption to limit our attention to only the plausible repair-limit replacement policies:

Assumption A.1: $TP_L(t_0) \le 0$ for all $t_0(\le 0)$.

Differentiating $TP_L(t_0)$ with respect to t_0 yields $dTP_L(t_0)/dt_0 = g(t_0)q_L(t_0)/T_L(t_0)^2$, where

$$q_L(t_0) = \{c + kL - (e_1 + k)t_0\}T_L(t_0) - (t_0 - L)V_L(t_0). \tag{5}$$

Then, we have the following result on the optimal repair-time limit under the assumption $V_L(t_0) \le 0$ for all t_0.

Theorem 1. *There exists a finite and unique optimal repair-time limit $t_0^*(0 < t_0^* < \infty)$ satisfying $q_L(t_0^*) = 0$ under the assumption (A.1). Then, the corresponding maximum long-run profit rate is given by*

$$TP_L(t_0^*) = \frac{c + kL - (e_1 + k)t_0^*}{t_0^* - L}. \tag{6}$$

3.2. Expected total discounted profit

Next, consider the case where the profit is discounted with the discount rate $\beta(> 0)$ over an infinite time horizon. Then, the expected total discounted profit during one cycle is given by

$$V_D(t_0) = \int_0^\infty \int_0^t e_0 \exp\{-\beta x\} dx dF(t) - \int_0^\infty \int_{t_0}^\infty c \exp\{-\beta(x+L)\} dG(t) dF(x)$$
$$- \int_0^\infty \int_0^{t_0} \int_0^t (e_1 + k) \exp\{-\beta(x+y)\} dy dG(t) dF(x)$$
$$- \int_0^\infty \int_{t_0}^\infty \int_0^L k \exp\{-\beta(x+y)\} dy dG(t) dF(x). \tag{7}$$

Since the expected present value of the unit profit just after one cycle is given by

$$\delta(t_0) = \int_0^\infty \int_0^{t_0} \exp\{-\beta(t+x)\} dG(t) dF(x)$$
$$+ \int_0^\infty \int_{t_0}^\infty \exp\{-\beta(x+L)\} dG(t) dF(x), \tag{8}$$

the expected total discounted profit over an infinite time horizon becomes

$$TP_D(t_0) = \sum_{j=0}^\infty V_D(t_0) \delta(t_0)^j = \frac{V_D(t_0)}{\bar{\delta}(t_0)}. \tag{9}$$

It is evident that

$$TP_L(t_0) = \lim_{\beta \to 0} \beta \cdot TP_D(t_0). \tag{10}$$

Of our interest is the derivation of the optimal repair-time limit $t_0^* \in [0, \infty)$ satisfying

$$TP_D(t_0^*) = \max_{0 \le t_0 < \infty} TP_D(t_0). \tag{11}$$

Similar to the long-run average profit, we make the following assumption:

Assumption A.2: $TP_D(t_0) \geq 0$ for all $t_0 (\geq 0)$.

Taking the differentiating of $TP_D(t_0)$ with respect to t_0, we have

$$\frac{d}{dt_0} TP_D(t_0) = \frac{g(t_0)\mathcal{L}\{f(\beta)\}}{\{\bar{\delta}(t_0)\}^2} q_D(t_0), \qquad (12)$$

where

$$q_D(t_0) = \left\{ \frac{-(k+e_1)(1-\exp\{-\beta t_0\})}{\beta} + k\int_0^L \exp\{-\beta y\}dy + c\exp\{-\beta L\} \right\} \bar{\delta}(t_0)$$

$$- \Big(\exp\{-\beta L\} - \exp\{-\beta t_0\}\Big) V_D(t_0). \qquad (13)$$

Then, we have the following theorem to characterize the optimal repair-time limit $t_0^* \in [0, \infty)$ satisfying

$$TP_D(t_0^*) = \max_{0 \leq t_0 < \infty} TP_D(t_0). \qquad (14)$$

Theorem 2. *Under the assumption (A.2), (i) If $q_D(0) > 0$ and $q_D(\infty) < 0$, then there exists a finite and unique optimal repair-time limit $t_0^* (0 < t_0^* < \infty)$ satisfying $q_D(t_0^*) = 0$, and the corresponding expected total discounted profit over an infinite time horizon is given by*

$$TP_D(t_0^*) = \frac{(k+e_1)(1-\exp\{-\beta t_0^*\}) - k(1-\exp\{-\beta L\}) - c\beta \exp\{-\beta L\}}{\beta(\exp\{-\beta t_0^*\} - \exp\{-\beta L\})}. \qquad (15)$$

(ii) If $q_D(0) \leq 0$, then the optimal repair-time limit is $t_0^* = 0$ and $TP_D(t_0^*) = TP_D(0)$.
(iii) If $q_D(\infty) \geq 0$, then the optimal repair-time limit is $t_0^* \to \infty$ and $TP_D(t_0^*) = TP_D(\infty)$.

In this section, we derived the optimal repair-time limit replacement policies maximizing two kinds of expected profit functions. It should be noted that the repair-time distribution has to be completely known, if one calculates the optimal repair-time limits according to Theorem 1 and Theorem 2. In general, however, it is not so easy to identify the repair-time distribution for a highly reliable equipment. In the following section, we will develop nonparametric algorithms to estimate the optimal repair-time limits under respective profit functions.

4. Graphical Methods

4.1. *Long-run average profit*

For a continuous repair time c.d.f., $p = G(t_0)$, define the Lorenz transform:

$$\phi(p) = \lambda \int_0^{G^{-1}(p)} t\, dG(t), \tag{16}$$

where

$$G^{-1}(p) = \inf\{t \geq 0 : G(t) \geq p\}, \tag{17}$$

if the inverse function exists, and $1/\lambda = \int_0^{G^{-1}(1)} t\, dG(t)$. Then, the curve $\mathcal{L} = (p, \phi(p)) \in [0,1] \times [0,1]$ is called the Lorenz curve. From a few algebraic manipulations, we obtain the following useful result to interpret the underlying optimization problem $\max_{0 \leq n_0 < \infty} TP_L(t_0)$ geometrically.

Theorem 3. *Under the assumption (A.1), if $c - e_1 L > 0$, then obtaining the optimal repair-time limit t_0^* maximizing the long-run average profit rate $TP_L(t_0)$ is equivalent to obtaining p^* $(0 \leq p^* \leq 1)$ such as*

$$\min_{0 \leq p \leq 1} \frac{\phi(p) + \beta_L}{p + \alpha_L}, \tag{18}$$

otherwise,

$$\max_{0 \leq p \leq 1} \frac{\phi(p) + \beta_L}{p + \alpha_L}, \tag{19}$$

where

$$\alpha_L = -\left\{1 + \frac{e_0 + e_1 + k}{\mu(e_1 L - c)}\right\}, \tag{20}$$

$$\beta_L = -\frac{\lambda}{\mu}\left\{\frac{(e_0 + k)L + c}{e_1 L - c}\right\}. \tag{21}$$

In fact, Theorem 3 is the dual of Theorem 1. From this result, it can be seen that the optimal repair-time limit $t_0^* = G^{-1}(p^*)$ is determined by calculating the optimal point $p^* (0 \leq p^* \leq 1)$ minimizing or maximizing the tangent slope from the point $(-\alpha_L, -\beta_L)$ to the curve $(p, \phi(p)) \in [0,1] \times [0,1]$.

4.2. Expected total discounted profit

For the discounted case, define the modified Lorenz transform:

$$\phi_\beta(p) \equiv \frac{\int_0^{G^{-1}(p)} \exp\{-\beta t\} dG(t)}{\int_0^{G^{-1}(1)} \exp\{-\beta t\} dG(t)}. \tag{22}$$

Theorem 4. *Under the assumption (A.2), if $(c\beta + +e_1)\exp\{-\beta L\} - e_1 > 0$, then obtaining the optimal repair-time limit t_0^* maximizing the expected total discounted profit over an infinite time horizon $TP_D(t_0)$ is equivalent to obtaining $p^* (0 \leq p^* \leq 1)$ such as*

$$\max_{0 \leq p \leq 1} : \frac{\phi_\beta(p) + \beta_D}{p + \alpha_D}, \tag{23}$$

otherwise

$$\min_{0 \leq p \leq 1} : \frac{\phi_\beta(p) + \beta_D}{p + \alpha_D}, \tag{24}$$

where

$$\alpha_D = \left(\frac{1}{\mathcal{L}\{f(\beta)\}} - 1\right)(k + e_0 + e_1) \Big/ \left[(c\beta + e_1)\exp\{-\beta L\} - e_1\right] - 1, \tag{25}$$

$$\beta_D = \Big[e_1 - [\mathcal{L}\{f(\beta)\}](k + e_0 + e_1) + (c\beta - k - e_0)]\exp\{-\beta L\}\Big]$$
$$\Big/ \Big[\mathcal{L}\{g(\beta)\}\mathcal{L}\{f(\beta)\}[(c\beta + e_1) + \exp\{-\beta L\} - e_1]\Big]. \tag{26}$$

5. Non-Parametric Estimation

Next, suppose that the optimal repair-time limit has to be estimated from n ordered complete observations: $0 = x_0 \leq x_1 \leq x_2 \leq \cdots \leq x_n$ of the repair times from a continuous c.d.f. $G(t)$, which is unknown. Then, the empirical distribution for this sample, is given by

$$G_{in}(x) = \begin{cases} i/n & \text{for } x_i \leq x < x_{i+1}, \\ 1 & \text{for } x_n \leq x, \end{cases} \tag{27}$$

where $i = 0, 1, 2, \ldots, n-1$. Then the Lorenz statistics [9], [1] can be defined by

$$\phi_{in} = \sum_{i=1}^{[np]} x_i \Big/ \sum_{i=1}^{n} x_i. \tag{28}$$

Theorem 5. *Define the estimator of the optimal repair-time limit by $\hat{t}_0^* = x_i^*$. Under the assumption (A.1), if $c > e_1 L$, an estimate of the optimal repair-time limit is given by*

$$\left\{ x_i^* \Big| \min_{0 \le i \le n} \frac{\phi_{in} + \beta_1}{i/n + \alpha_1} \right\}, \tag{29}$$

otherwise,

$$\left\{ x_i^* \Big| \max_{0 \le i \le n} \frac{\phi_{in} + \beta_1}{i/n + \alpha_1} \right\}. \tag{30}$$

In fact, Goldie [9] proved the strong consistency of the Lorenz statistics given in Eq.(28), that is, $\phi_{in} \to \phi(p)$ as $n \to \infty$ a.s. This fact means that the estimator of the optimal repair-time limit $\hat{t}_0^*$ may be also consistent. The graphical procedure proposed here has an educational value for better understanding of the optimization problem and it is convenient for performing sensitivity analysis of the optimal repair-time limit when different values are assigned to the model parameters. The special interest is, of course, to estimate the optimal repair-time limit without specifying the repair time distribution. Although some theoretical distribution functions such as the lognormal distribution are often assumed for the repair time distribution, our non-parametric estimation algorithm can estimate the optimal repair-time limit based on the on-line knowledge about the observed repair times.

Next, define the statistics of the modified Lorenz transform [4] by

$$\phi_{in}^\beta = \left[1 - \left(1 - \frac{i}{n}\right) \exp\{-\beta x_i\} - \beta \sum_{j=1}^{i} \left(1 - \frac{j-1}{n}\right) (x_j - x_{j-1}) \exp\{-\beta x_j\} \right]$$

$$\Big/ \left[1 - \beta \sum_{j=1}^{n} \left(1 - \frac{j-1}{n}\right) (x_j - x_{j-1}) \exp\{-\beta x_j\} \right]. \tag{31}$$

By plotting $(i/n, \phi_{in}^\beta)$ and connecting them by line segment, we obtain the modified sample Lorenz curve.

Theorem 6. *Define the estimator of the optimal repair-time limit by $\hat{t}_0^* = x_i^*$. Under the assumption (A.2), if $(c\beta + e_1) \exp\{-\beta L\} - e_1 > 0$, then*

$$\left\{ x_i^* \Big| \max_{0 \le i \le n} \frac{\phi_{in}^\beta + \beta_D}{i/n + \alpha_D} \right\}, \tag{32}$$

otherwise

$$\left\{ x_i^* \Big| \min_{0 \le i \le n} \frac{\phi_{in}^\beta + \beta_D}{i/n + \alpha_D} \right\}. \tag{33}$$

In this paper, we have developed estimation algorithms for the optimal repair-time limits under different earning rate criteria. As mentioned before, it can be easily understood that the estimator of the optimal repair-time limit maximizing the long-run average profit rate has strongly consistent, i.e., $\hat{t}_0^* \to t_0^*$ as $n \to \infty$. However, it is an open question whether the estimate in the discounted case can converge to the real optimum as $n \to \infty$. For the practical purpose, the convergence properties of nonparametric estimators for the optimal repair-time limits should be checked through a simulation study.

Acknowledgements: This research was partially supported by the Ministry of Education, Science, Sports and Culture, Grant-in-Aid for Scientific Research (B), Grant No. 1631011600 (2004-2006) and Scientific Research (C), Grant No. 18510138 (2006-2008).

References

1. Chandra, M. and Singpurwalla, N. D. (1981), Relationship between some notions which are common to reliability and economics, *Math. Ope. Res.*, **6**, 113–121.
2. Dohi, T., Aoki, T., Kaio, N. and Osaki, S. (1998), Nonparametric preventive maintenance optimization models under earning rate criteria, *IIE Trans. Qual. Reliab. Eng.*, **30**, 1099–1108 (1998).
3. Dohi, T., Ashioka, A., Kaio, N. and Osaki, S. (2003), The optimal repair-time limit replacement policy with imperfect repair: Lorenz transform approach, *Math. Computer Modelling*, **38**, 1169–1176.
4. Dohi, T., Ashioka, A., Kaio, N. and Osaki, S. (2006), Statistical estimation algorithms for some repair-limit replacement scheduling problems under earning rate criteria, *Computers & Math. Applic.*, **51**, 345–356.
5. Dohi, T., Kaio, N. and Osaki, S. (2001), Total time on test processes and their application to maintenance problem, *System and Bayesian Reliability - Essays in Honor of Profesor Richard E. Barlow on His 70th Birthday* (Y. Hayakawa, T. Irony and M. Xie, Eds.), 123–143, World Scientific, Singapore.
6. Dohi, T. and Kaio, N. (2005), Repair-limit replacement program in industrial maintenance - renewal reward process modeling, *Applied Economic Informatics and Systems Sciences* (T. Tokimasa, S. Hiraki and N. Kaio, Eds.), 157–172, Kyushu University Press, Fukuoka.
7. Dohi, T., Takeita, K. and Osaki, S. (2000), Graphical methods for determining/estimating optimal repair-limit replacement policies, *Int. J. Reli., Qual. Safe. Eng.*, **7**, 43–60.
8. Gastwirth, J. L. (1971), A general definition of the Lorenz curve, *Econometrica*, **39**, 1037–1039.
9. Goldie, C. M. (1977), Convergence theorems for empirical Lorenz curves and their inverses, *Adv. Appl. Probab.*, **9**, 765–791.

10. Hastings, N. A. J. (1969), The repair limit replacement method, *Opel. Res. Quart.*, **20**, 337–349.
11. Lorenz, M. O. (1893), Methods of measuring the concentration of wealth, *J. Amer. Statist. Assoc.*, **9**, 209–219.
12. Love, C. E. and Guo, R. (1996), Utilizing Weibull failure rates in repair limit analysis for equipment replacement/preventive maintenance decisions, *J. Opel. Res. Soc.*, **47**, 1366–1376.
13. Nakagawa, T. (1977), Optimum preventive maintenance and repair limit policies maximizing the expected earning rate, *R.A.I.R.O.-Ope. Res.*, **11**, 103–108.
14. Nguyen, D. G. and Murthy, D. N. P. (1980), A note on the repair limit replacement policy, *J. Opel. Res. Soc.*, **31**, 1103–1104.

INVESTIGATION OF EQUIVALENT STEP-STRESS TESTING PLANS

E. A. ELSAYED, Y. ZHU AND H. ZHANG[†]

Department of Industrial and System Engineering, Rutgers, The State University of New Jersey, 96 Frelinghuysen Road, Piscataway, NJ 08854-8018, USA
elsayed@rci.rutgers.edu

H. LIAO

Industrial and Manufacturing Engineering Department, Wichita State University KS, 67260, USA

Accelerated life testing (ALT) is a method for obtaining failure time data of test units quickly under more severe conditions than the normal operating conditions. Typical ALT plans require the determination of stress types, stress levels, allocation of test units to those stress levels, duration of the test and other test parameters. Traditionally, ALT is conducted under constant stresses during the entire test duration. In practice, the constant-stress tests need more test unites and a long time at low stress levels to yield sufficient failure data. However, due to budget and time constraints, there are increasing necessities to design testing plans that can shorten the test duration and reduce the total cost while achieving the equivalent accuracy of reliability estimate. In this chapter, we develop an equivalent step-stress testing plan such that the reliability predictions at normal conditions using the results of this plan will be approximately "equivalent" to the corresponding constant-stress test plan but the test duration is significantly shortened. We determine the optimum parameters of the test plan through a numerical example and evaluate the equivalence of the test plans using simulation. We also investigate the sensitivity of the ALT model parameters.

1. Introduction

The significant increase in the introduction of new products coupled with the significant reduction in time from product design to manufacturing, as well as the increasing customer's expectation for high reliability, have prompted industry to shorten its product test duration. In many cases, ALT might be the only feasible approach to meet this requirement. The accuracy of the statistical inference procedure obtained using ALT data has a profound effect on the reliability estimates and the subsequent decisions regarding system configuration, warranties and preventive maintenance schedules. Specifically,

---

[†] This work is supported by the National Science Foundation Award No. DMI-0619991.

the reliability estimate depends on two factors, the ALT model and the experimental design of test plans. Without an optimal test plan, it is likely that a sequence of expensive and time-consuming tests results in inaccurate reliability estimates and improper final product design requirements. That might also cause delays in product release, or the termination of the entire product development.

Traditionally, ALT is conducted under constant stresses during the entire test. For example, a typical constant temperature test plan consists of defining three temperature levels (high, medium, and low) and test units are divided among these three levels where units at each level are exposed to the same temperature until failure or until the test is terminated. The test results are then used to extrapolate the product life at normal conditions. In practice, constant-stress tests are easy to carry out but need more test units and long time at the low stress level to yield "enough" number of failures. However, in many cases the available number of test units and test duration are extremely limited. This has prompted industry to consider step-stress tests where the test units are first subjected to a lower stress level for some time; if no failures or only a small number of failures occur, the stress is increased to a higher level and held constant for another amount of time; the steps are repeated until all units fail or the predetermined test time has expired.

Usually, step-stress tests yield failures in a much shorter time than constant-stress tests, but the statistical inference from the data is more difficult to make. Moreover, since the test duration is short and a large proportion of failures occur at high stress levels far from the design stress level, much extrapolation has to be made, which may lead to poor estimation accuracy. On the other hand, there could be other choices in stress loadings (e.g., cyclic-stress and ramp-stress) in conducting ALT experiments. Each stress loading has some advantages and drawbacks. This has raised many practical questions such as: Can accelerating test plans involving different stress loadings be designed such that they are equivalent? What are the measures of equivalency? Can such test plans and their equivalency be developed for multiple stresses especially in the setting of step-stress tests and other profiled stress tests? When and in which order should we change the stress levels in multi-stress multi-step tests?

Figure 1 shows various stress loading types as well as their adjustable parameters. These stress loadings have been widely utilized in ALT experiments. For instance, static-fatigue tests and cyclic-fatigue tests [1] have been frequently performed on optical fibers to study their reliability; dielectric-breakdown of thermal oxides [2] have been studied under elevated constant electrical fields and temperatures; the lifetime of ceramic components subject to slow crack growth due to stress corrosion have been investigated under cyclic stress by Choi and Salem [3].

Investigation of Equivalent Step-Stress Testing Plans 153

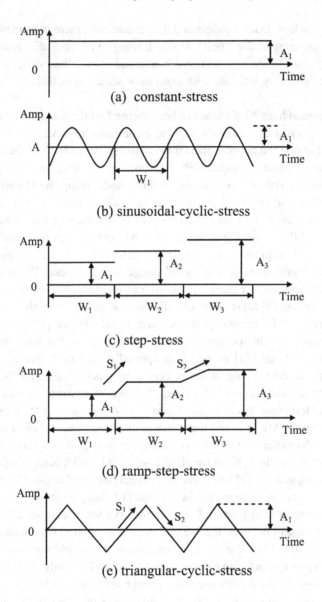

Figure 1. Various stress loading types.

These types of stress loading are selected by practitioners based on the simplicity of associated statistical analyses, and familiarity with existing analytical tools and industrial routines without following a systematic refinement procedure. Due to budget and time constraints, there is an increasing necessity to determine the "best" stress loading type and the associated parameters in order to shorten the test duration and reduce the total cost while achieving the reliability estimate with equivalent accuracy to that of constant-stress testing.

Current research on ALT Plans has been focused on the design of optimum test plans for given stress loading type. Under constant-stress ALT, Nelson and Kielpinski [4] obtain the optimal plans for estimating the medians of the normal and lognormal distributions respectively. Maxim et al. [5] determine the optimal ALT plans for bivariate exponential or Weibull model using the D-optimality criterion. Meeker and Hahn [6] consider the optimal allocation of test units to accelerated stress conditions with the objective of minimizing the estimate of the product reliability under the design stress conditions. Nelson and Meeker [7] provide optimal test plans that estimate the percentiles of the Weibull and extreme-value distributions at a specified design stress. Meeker [8] compares the statistically optimal test plans to a more practical test plan involving three stress levels. Nelson [9] reviews statistically optimal and compromise plans for the single stress ALT planning problem. Meeter and Meeker [10] extend the existing maximum likelihood theory for planning ALT to the variable scale parameter model. Yang [11] proposes an optimal design of 4-level constant-stress ALT plans considering different censoring times. Regarding the step-stress ALT, Miller and Nelson [12] present the cumulative exposure model and obtain optimal test plans that minimize the asymptotic variance of the maximum likelihood estimate (MLE) of the mean life at the design stress. Bai et al. [13] extend the results to the case where a prescribed censoring time is involved. Bai and Chun [14] obtain the optimal simple step-stress ALT with competing causes of failure. Chung and Bai [15] consider the optimal design of step-stress ALT in which a cumulative exposure model for the effect of changing stress is assumed. Khamis and Higgins [16] present 3-step step-stress plans assuming a linear or quadratic relationship between the life and the stress. Xiong [17] addresses the effect of the statistical inferences on the parameters of a simple step-stress ALT model with Type II censoring. Xiong and Milliken [18] study the statistical models in step-stress ALT when the stress change times are random. Xiong and Ji [19] study the optimal design of a simple step-stress test plan involving grouped and censored data.

The wide range of stress applications, stress levels and corresponding test durations give rise to the equivalency between test plans. Interestingly, because of wide range of polymer based advanced composite materials that are used in certified aircraft applications as well as the large variability of properties among these composites and within the same batch of a composite material the manufacturers and the Federal Aviation Administration developed a procedure to assess equivalence between different polymer based composites. The procedure utilizes essentially small data sets to generate test condition statistics such as population variability and corresponding basis values to pool results for a specific failure mode across all environments. The statistics from the test are compared and assessment of the "equivalency" is then made based on the mean and variance of the data, Tomblin et al. [20]. Clearly, the term equivalency here refers to basic statistics about samples from populations but it does not provide information on reliability prediction or other time-dependent characteristics. A brief literature review shows that fundamental research on the equivalency of test plans has not yet been investigated in the reliability engineering. Without the understanding of such equivalency, it is difficult for a test engineer to determine the best experimental settings before conducting actual ALT.

Furthermore, as is often the case, products are usually exposed to multiple stress types in actual use such as temperature, humidity, electric current, and various types of vibration. To study the reliability of such products, it is important to subject test units to multiple stress types simultaneously in ALT experiments. For constant-stress tests, it might not be difficult to extend the statistical methods in the design of optimum test plans for single stress to multiple stresses scenarios. However, the problem becomes complicated when time-varying stresses such as step-stresses are considered. For example, in a multiple-stress-type multi-step test, issues such as when and in which order the levels of the stresses should be changed are challenging and unsolved. Figure 2 illustrates two experimental settings out of thousands of choices as one can imagine in conducting a multiple-stress-type multi-step ALT.

In general, an arbitrary selection from combinations of multiple stress profiles may not result in the most accurate reliability estimates, especially when the effects of the stresses on the reliability of the product are highly correlated. Therefore, optimization of test plans by tuning the high dimensional decision variables under time and cost constraints needs to be carefully investigated from the perspective of statistics, operations research and engineering physics.

2. Definitions of Equivalency

We consider different optimization criteria for test plans equivalence depending on the type of stress loading and the objective of the test. One of the optimization criteria to be considered is the minimization of the asymptotic variance of the maximum likelihood estimate (MLE) of the mean time to failure at normal operating conditions.

To study the equivalency among ALT plans involving different stress loadings, several definitions are explored. Some definitions are:

Definition 1. Two ALT plans are equivalent if they generate the same values of the optimization criterion under Type I censoring.

Definition 2. Two ALT plans are equivalent if the difference between the estimated times to failure and the respective confidence intervals by the two plans at normal operating conditions are within $\delta\%$, where δ is an acceptable level of deviation.

In the following sections, we investigate the equivalent ALT plans based on the first definition.

3. Equivalency of Step-Stress and Constant-Stress ALT Plans

In this chapter, we consider the case of the equivalency of step-stress ALT plans and constant-stress ALT plans. Since constant-stress tests are the most commonly conducted accelerated life tests in industry and their statistical inference has been extensively investigated, we focus on determining the equivalent step-stress ALT plan to the constant stress ALT plan. The constant-stress ALT plan serves as the baseline test result for comparison with the step-stress plan. More importantly, the constant-stress ALT plan requires longer test duration when compared with other test plans. Therefore, the efficiency of equivalent plans will also be measured by the percent reduction in test time. In this section, we are interested in determining the minimum test duration of the step-stress ALT plan, which is equivalent to the baseline constant-stress ALT plan.

To analyze the failure time data from a step stress test, we relate the life distribution under step-stresses to the distribution under constant stresses. The cumulative exposure (CE) model proposed by Nelson [21] has been widely used as an effective approach for this problem. This model assumes that:

- The remaining lifetime of a test unit depends on the current cumulative fraction failed and the current stress regardless how the fraction is accumulated.
- If held at the current stress, survivors fail according to the cumulative distribution for that stress, but starting at the previously accumulated fraction failed.

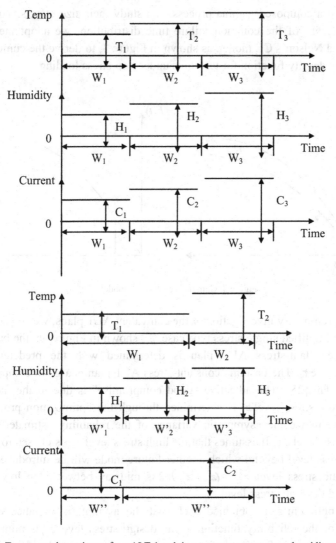

Figure 2. Two example settings of an ALT involving temperature, current, humidity and electric current.

Hirose [22] proposes a generalized Weibull CE model with a threshold parameter. Xiong and Milliken [18] study exponential CE models with random stress-change times ALT. Komori [23] constructs a generalized CE model having a threshold parameter that decides whether a specimen is influenced by stress or not. Ballani et al. [24] create two CE models, the inverse gamma process and a composed gamma process, and study their size effects. To relax the requirement of the common failure time distribution, we adopt the most widely used Nelson's CE model, as shown in Figure 3, to derive the cumulative failure time density function for test units under step-stress loading.

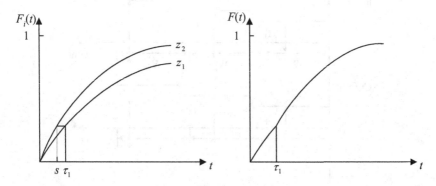

Figure 3. Cumulative exposure model.

As a preliminary investigation of the equivalent ALT plans, we consider the following simplified single stress type case. As shown in Figure 4a, the baseline optimum constant-stress ALT plan is determined with the predetermined censoring time τ. The baseline constant-stress ALT plan is a good compromise optimum plan [25]. The adjective "good compromise" is due to the fact that only the low stress level z_L is determined through the optimization process in order to minimize the asymptotic variance of the reliability estimates at the design stress level z_D. It assumes that the high stress level z_H is chosen to be the highest stress level beyond which another failure mode will be introduced. The intermediate stress level $z_M = (z_L + z_H)/2$ is midway between the low stress level z_L and the high stress level z_H.

The optimum z_L is obtained such that the asymptotic variance of ML estimates of the reliability function at the design stress level z_D is minimized. The allocation of fraction of test units to the stress levels z_L, z_M and z_H are $p_L = 4/7$, $p_M = 2/7$, and $p_H = 1/7$ respectively. This unequal allocation is a compromise that extrapolates reasonably well. For a sample of n test units, the allocation is, $n_L = 4n/7$, $n_M = 2n/7$, and $n_H = n/7$.

Assuming that the proportional odds (PO) model is assumed to fit the failure time data which is expressed as $\theta(t;z) = \exp(\boldsymbol{\beta}'z)\theta_0(t)$, where $z = (z_1, z_2, \ldots, z_k)'$ is a column vector of stress levels; $\boldsymbol{\beta} = (\beta_1, \beta_2, \ldots, \beta_k)'$ is a

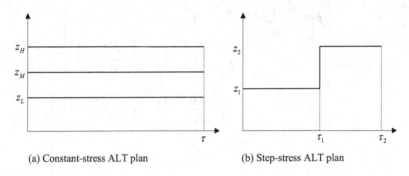

(a) Constant-stress ALT plan   (b) Step-stress ALT plan

Figure 4. An example of equivalent constant ALT plan and step-stress ALT plan.

column vector of model parameters and baseline odds function $\theta_0(t) = \gamma_1 t + \gamma_2 t^2$, where γ_1 and γ_2 are unknown parameters, there is no intercept term because the odds function always crosses the origin [26]. Therefore the optimization problem can be formulated as follows:

Objective function
Min

$$f_C(x) = \int_0^T Var[\hat{R}_C(t;z_D)]dt \qquad (1)$$

Subject to

$$\Sigma_C = F_C^{-1} \qquad (2)$$

$$p_L = 4/7, \ p_M = 2/7, \ p_H = 1/7 \qquad (3)$$

$$z_D \leq z_L, \ z_H = z_{upper}, \ z_M = (z_L + z_H)/2 \qquad (4)$$

$$np_L[1 - R(\tau; z_L)] \geq MNF \qquad (5)$$

$$np_M[1 - R(\tau; z_M)] \geq MNF \qquad (6)$$

$$np_H[1 - R(\tau; z_H)] \geq MNF \qquad (7)$$

where F_c is the Fisher Information matrix for the constant stress test plan and is obtained as

$$F_C = n_L \begin{bmatrix} E\left[-\dfrac{\partial^2 l_C(z_L)}{\partial \beta^2}\right] & 0 & 0 \\ 0 & E\left[-\dfrac{\partial^2 l_C(z_L)}{\partial \gamma_1^2}\right] & E\left[-\dfrac{\partial^2 l_C(z_L)}{\partial \gamma_1 \partial \gamma_2}\right] \\ 0 & E\left[-\dfrac{\partial^2 l_C(z_L)}{\partial \gamma_1 \partial \gamma_2}\right] & E\left[-\dfrac{\partial^2 l_C(z_L)}{\partial \gamma_2^2}\right] \end{bmatrix}$$

$$+ n_M \begin{bmatrix} E\left[-\dfrac{\partial^2 l_C(z_M)}{\partial \beta^2}\right] & 0 & 0 \\ 0 & E\left[-\dfrac{\partial^2 l_C(z_M)}{\partial \gamma_1^2}\right] & E\left[-\dfrac{\partial^2 l_C(z_M)}{\partial \gamma_1 \partial \gamma_2}\right] \\ 0 & E\left[-\dfrac{\partial^2 l_C(z_M)}{\partial \gamma_1 \partial \gamma_2}\right] & E\left[-\dfrac{\partial^2 l_C(z_M)}{\partial \gamma_2^2}\right] \end{bmatrix}$$

$$+ n_H \begin{bmatrix} E\left[-\dfrac{\partial^2 l_C(z_H)}{\partial \beta^2}\right] & 0 & 0 \\ 0 & E\left[-\dfrac{\partial^2 l_C(z_H)}{\partial \gamma_1^2}\right] & E\left[-\dfrac{\partial^2 l_C(z_H)}{\partial \gamma_1 \partial \gamma_2}\right] \\ 0 & E\left[-\dfrac{\partial^2 l_C(z_H)}{\partial \gamma_1 \partial \gamma_2}\right] & E\left[-\dfrac{\partial^2 l_C(z_H)}{\partial \gamma_2^2}\right] \end{bmatrix} \quad (8)$$

$$x = z_L \quad (9)$$

The only decision variable in the above optimization problem is the low stress level z_L. Let z_L^* denote the optimum solution, and $f_C(z_L^*)$ denote the minimum asymptotic variance of the reliability estimates given by the constant-stress ALT plan. Compared with the constant-stress ALT plan, the step-stress ALT plan can substantially shorten the test duration. The censoring time τ_2 of the step-stress ALT plan shown in Figure 4b is less than the censoring time τ of the constant-stress ALT plan shown in Figure 4a. The objective of the equivalency of ALT plans is to determine the minimum τ_2 which results in the equivalent asymptotic variances of the reliability estimates at the design stress level obtained from the two test plans.

To obtain the optimum asymptotic variance of the reliability estimates at the design stress level for any given τ_2 of the step-stress ALT plan, assuming that the PO model holds we formulate the optimization problem as follows:

Objective function
Min

$$f_S(\mathbf{x}) = \int_0^T Var[\hat{R}_S(t;z_D)]dt \tag{10}$$

Subject to

$$n\Pr[t \leq \tau_1; z_1] \geq MNF_1 \tag{11}$$

$$n\Pr[t \leq \tau_2; z_1, z_2] \geq MNF_2 \tag{12}$$

$$z_2 = z_{upper} \tag{13}$$

$$\Sigma_S = F_S^{-1} \tag{14}$$

where

$$\mathbf{x} = \begin{bmatrix} z_1 \\ \tau_1 \end{bmatrix} \tag{15}$$

F_S is the Fisher Information matrix for the step-stress test plan

$$F_S = n \begin{bmatrix} E\left[-\dfrac{\partial^2 l_S}{\partial \beta^2}\right] & 0 & 0 \\ 0 & E\left[-\dfrac{\partial^2 l_S}{\partial \gamma_1^2}\right] & E\left[-\dfrac{\partial^2 l_S}{\partial \gamma_1 \partial \gamma_2}\right] \\ 0 & E\left[-\dfrac{\partial^2 l_S}{\partial \gamma_1 \partial \gamma_2}\right] & E\left[-\dfrac{\partial^2 l_S}{\partial \gamma_2^2}\right] \end{bmatrix} \tag{16}$$

MNF_1 and MNF_2 are the minimum required number of failures at low stress level and high stress level respectively.

Solving the above optimization problem, we obtain the optimum solution: z_1^*, τ_1^*, and the achieved minimized asymptotic variance $f_S(z_1^*, \tau_1^*; \tau_2)$ for given τ_2. Following the definition of the equivalent

ALT plans in section 2, the minimum censoring time τ_2^* of the step-stress ALT plan is defined as:

$$\tau_2^* = \inf\{\tau_2 \mid f_S(z_1^*, \tau_1^*; \tau_2) \le f_C(z_L^*)\} . \qquad (17)$$

We use the following bisectional search procedure to determine the minimum censoring time τ_2^*:

1. Step 1. Solve the nonlinear optimization problem defined in the constant-stress ALT plan, and find $f_C(z_L^*)$, choose a small value ε arbitrarily.
2. Step 2. Let $i = 1$, $\tau_L = 0$, $\tau_H = \tau$, and $\tau_2^i = (\tau_L + \tau_H) = \tau/2$.
3. Step 3. Determine $f_S(z_1^*, \tau_1^*; \tau_2^i)$ by solving the above nonlinear optimization problem defined in the step-stress ALT plan.
4. Step 4. Let $i = i+1$. If $f_S(z_1^*, \tau_1^*; \tau_2^{i-1}) \le f_C(z_L^*)$, then let $\tau_L = \tau_2^{i-1}$ and $\tau_2^i = (\tau_L + \tau_H)/2$; otherwise let $\tau_H = \tau_2^{i-1}$ and $\tau_2^i = (\tau_L + \tau_H)/2$.
5. Step 5. Determine $f_S(z_1^*, \tau_1^*; \tau_2^i)$ by solving the above nonlinear optimization problem defined in the step-stress ALT plan. If $|\tau_2^i - \tau_2^{i-1}| \le \varepsilon$ and $f_S(z_1^*, \tau_1^*; \tau_2^i) \le f_C(z_L^*)$, then stop and let $\tau_2^* = \tau_2^i$; otherwise continue Step 4.

4. Numerical Example and Simulation Study

4.1. Numerical example

A constant-stress accelerated life test is conducted at three temperature levels for MOS devices in order to estimate its reliability function at design temperature level of 25°C. The test needs to be completed in 100 hours. The total number of available test units is 200. The highest temperature level that test units can experience is 250°C. It is expected that the ALT will provide the most accurate reliability estimate over a period of 10000 hours. A constant-stress ALT plan can be designed through the following steps:

1. According to Arrhenius model, the scaled stress $z = 1000/(temp + 276.13)$ is used. Then the design stress level $z_D = 3.35$ and the upper bound are $z_{upper} = 1.19$.
2. A baseline experiment is conducted to obtain a set of initial values of the parameters for the PO model with quadratic odds function [26]. These values are: $\beta = -1.8$, $\gamma_1 = 10$, and $\gamma_2 = 0.001$.

3. Following the formulation in section 3 for the constant-stress ALT plan, the nonlinear optimization problem is expressed as follows:

Objective function
Min

$$f_C(x) = \int_0^{10000} Var[\hat{R}_C(t;3.35)]dt \tag{18}$$

Subject to

$$\Sigma_C = F_C^{-1} \tag{19}$$

$$p_L = 4/7, \; p_M = 2/7, \; p_H = 1/7 \tag{20}$$

$$3.35 = z_D \leq z_L, \; z_H = z_{upper} = 1.91, \; z_M = (z_L + z_H)/2 \tag{21}$$

$$np_L[1 - R(\tau; z_L)] \geq 90 \tag{22}$$

$$np_M[1 - R(\tau; z_M)] \geq 50 \tag{23}$$

$$np_L[1 - R(\tau; z_H)] \geq 40 \tag{24}$$

where F_c is the Fisher Information matrix for the constant stress test plan which is defined earlier.

$$n = 200, \; \tau = 100 \quad x = z_L \tag{25}$$

4. The solution of the optimization problem is: $z_L^* = 3.1$ and $f_C(3.1) = 0.2823$.

In order to shorten the test period and still obtain the equivalent asymptotic variance, we consider the step-stress ALT plan. The nonlinear optimization problem of the step-stress ALT plan is formulated as follows:

Objective function
Min

$$f_S(x) = \int_0^{10000} Var[\hat{R}_S(t;3.35)]dt \tag{26}$$

Subject to

$$n\Pr[t \leq \tau_1; z_1] \geq 50 \tag{27}$$

$$n\Pr[t \le \tau_2; z_1, z_2] \ge 140 \tag{28}$$

$$z_2 = 1.19 \tag{29}$$

$$\Sigma_S = F_S^{-1} \tag{30}$$

where

$$x = \begin{bmatrix} z_1 \\ \tau_1 \end{bmatrix} \tag{31}$$

$$F_S = n \begin{bmatrix} E\left[-\dfrac{\partial^2 l_S}{\partial \beta^2}\right] & 0 & 0 \\ 0 & E\left[-\dfrac{\partial^2 l_S}{\partial \gamma_1^2}\right] & E\left[-\dfrac{\partial^2 l_S}{\partial \gamma_1 \partial \gamma_2}\right] \\ 0 & E\left[-\dfrac{\partial^2 l_S}{\partial \gamma_1 \partial \gamma_2}\right] & E\left[-\dfrac{\partial^2 l_S}{\partial \gamma_2^2}\right] \end{bmatrix} \tag{32}$$

$$n = 200 \tag{33}$$

The minimum censoring time τ_2^* of the step-stress ALT plan is determined by:

$$\tau_2^* = \inf\{\tau_2 \mid f_S(z_1^*, \tau_1^*; \tau_2) \le f_C(z_L^*)\}. \tag{34}$$

Following the bisectional search algorithm (ε is arbitrarily chosen as 1) in Section 3, the results are listed in Table 1.

Table 1: Results of bisectional search.

i	τ_2^i	$f_S(z_1^*, \tau_1^*; \tau_2^i)$
1	50	0.2884
2	75	0.2489
3	63	0.2648
4	56	0.2764
5	53*	0.2822
6	55	0.2782
7	54	0.2802

From the above table, the minimum censoring time is $\tau_2^* = 53$. The corresponding optimum low stress level is $z_1 = 3.15$; the optimum stress changing time is $\tau_1 = 46$.

4.2. Sensitivity study

Table 2 presents sensitivity analysis of the minimum censoring time with step-stress loading when the estimation of the covariate effects β and PO model parameter γ_1 and γ_2 are not accurate. It can be seen that the minimum censoring time τ_2^*, the optimum stress changing time τ_1, and the optimum low stress level z_1 are insensitive to the PO model parameter γ_1 and γ_2, but are slightly sensitive to the deviation of covariate effects β. However, accurate estimates of these parameters can be achieved by well controlled experiments combined with corrected score [27] or corrected likelihood method [28].

Table 2: Results of bisectional search.

Deviation	β		γ_1		γ_2	
	-10%	10%	-10%	10%	-10%	10%
Minimum censoring time τ_2^*	40 (24.5%)	59 (35.8%)	54 (1.9%)	49 (7.5%)	51 (3.8%)	51 (3.8%)
Optimum stress changing time τ_1	36 (21.7%)	52 (13%)	48 (4.3%)	44 (4.3%)	45 (2.22%)	45 (2.22%)
Optimum low stress level z_1	3.3 (4.8%)	3.05 (3.2%)	3.1 (1.6%)	3.15 (1.6)	3.1 (1.6%)	3.1 (1.6%)

4.3. Simulation study

In this section, we conduct a simulation study to verify that the resultant constant ALT plan and step-stress ALT plan in the previous section are indeed equivalent in terms of the estimation accuracy.

Sets of simulation data following both constant ALT plan and step-stress ALT plan are generated by the Monte Carlo simulation method based on the following reliability function.

$$R(t) = \frac{1}{1+(\gamma_1 t)\exp(\beta z)} \tag{35}$$

It can be easily verified that the assumption of the PO model is valid for the failure time samples generated by the above reliability function. The corresponding PO model can be expressed as

$$\theta(t; z) = \theta_0(t)\exp(\beta z) = (\gamma_1 t)\exp(\beta z). \tag{36}$$

The total sample size is 200 and is divided into three stress groups with constant ALT plan: $n_{total} = 200$, $n_L = 800/7 \approx 114$, $n_M = 57$, and $n_H = 29$. The model parameters used to generate the Monte Carlo simulation failure times are summarized in Table 3 for constant ALT plan.

Table 3: Model parameters for constant-stress ALT simulation.

Common parameters	Censoring time: $\tau = 100$; Design stress level: $z_D = 3.5$; Baseline parameters: $\beta = -1.8$, $\gamma_1 = 10$		
Stress	Low	Medium	High
Stress level	$z_L = 3.1$	$z_M = 2.15$	$z_H = 1.19$
Sample size	$n_L = 114$	$n_M = 57$	$n_H = 29$
Dataset	I	II	III

Rewriting Eq. (35) by solving t, we obtain

$$t = \frac{F(t)\exp(-\beta z)}{\gamma_1[1-F(t)]}. \tag{37}$$

Therefore 114 simulated failure times are generated for low stress level by the following equation

$$t_i = \frac{rand(i)\exp(-\beta z_L)}{\gamma_1[1-rand(i)]}, \quad \text{for } i = 1, \ldots, 114 \tag{38}$$

where $rand(i)$ is uniformly distributed random variable on the interval (0, 1).

Similarly, 57 simulated failure times are generated for medium stress level based on the equation

$$t_i = \frac{rand(i)\exp(-\beta z_M)}{\gamma_1[1-rand(i)]}, \quad \text{for } i = 115, \ldots, 171. \qquad (39)$$

Finally, there are 29 simulated failure times generated for high stress level derived from the equation and simulated failure time greater than 100 is censored.

$$t_i = \frac{rand(i)\exp(-\beta z_H)}{\gamma_1[1-rand(i)]}, \quad \text{for } i = 172, \ldots, 200. \qquad (40)$$

Likewise, the model parameters for step-stress ALT simulation are listed in Table 4.

Table 4: Model parameters for step-stress ALT simulation.

Common parameters	Censoring time: $\tau_2 = 53$; Design stress level: $z_D = 3.5$; Stress changing time $\tau_1 = 46$; Baseline parameters: $\beta = -1.8$, $\gamma_1 = 10$	
Stress	Low	High
Stress level	$z_1 = 3.15$	$z_2 = 1.19$
Sample size	$n = 200$	
Dataset	VI	V

Using the model parameters in Table 4, the Monte Carlo simulation data for the step-stress ALT are generated based on the Nelson's CE model.

More specifically, the procedure of generation of the Monte Carlo simulation data for the step-stress ALT is

1. A total of 200 failure times is generated based on the equation

$$t_i = \frac{rand(i)\exp(-\beta z_1)}{\gamma_1[1-rand(i)]}, \quad \text{for } i = 1, \ldots, 200. \qquad (41)$$

2. Among the 200 failure times simulated in step 1, all failure times greater than τ_1 are eliminated.
3. Suppose that only n_1 failure times are kept in step 2. The remaining $200 - n_1$ failure time are generated based on the equation

$$t_i = \frac{rand(i)\exp(-\beta z_2)}{\gamma_1[1-rand(i)]} + \tau_1 - s, \quad \text{for } i = n_1+1,\ldots,200. \quad (42)$$

4. Among the $200-n_1$ failure times simulated in step 3, all failure times smaller than τ_1 are discarded, all failure times greater than τ_2 are censored.

The results of the reliability estimations from the simulated failure time data based on the constant-stress ALT plan and the step-stress ALT plan are shown in Figure 5. As shown in Figure 5, the estimated reliability functions are so close that the equivalency of the constant-stress ALT plan and step-stress ALT plan is verified.

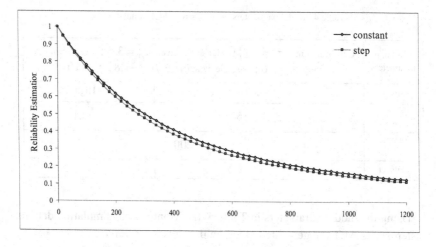

Figure 5. Estimated reliability functions from constant and step simulated data.

5. Conclusion and Future Work

In this chapter we present the equivalency of ALT plans involving different stress-loading types. The definitions of the equivalent ALT plans are discussed. The equivalency of constant-stress ALT plan and step-stress ALT plan is investigated for single stress type problem. Under this scenario, the baseline constant-stress ALT plan is given, the equivalent step-stress ALT plan is determined based on the definition of the equivalency of ALT plans.

The numerical example shows that the optimal step-stress ALT plan significantly reduces the test duration. Simulation study verifies the equivalency of the constant-stress ALT plan and step-stress ALT plan. The research of the equivalent ALT plans enables reliability practitioners to choose the appropriate ALT plan to accommodate any resource restrictions and duration of the test.

References

1. M. J. Mathewson and H.H. Yuce, Kinetics of degradation during fatigue and aging of fused silica optical fiber, In *Proceedings of SPIE* **2290**, 204, (1994).
2. E. A. Elsayed, H. T. Liao, and X. D. Wang, An extended linear hazard regression model with application to time-dependent-dielectric-breakdown of thermal oxides, *IIE Transactions on Quality and Reliability* **38(1)**, 1, (2006).
3. S. R. Choi and J. A. Salem, Error in flexure testing of advanced ceramics under cyclic loading, *Ceramic Engineering and Science Proceedings* **18** (3), 495, (1997).
4. W. B. Nelson and T. J. Kielpinski, Theory for optimum accelerated life tests for normal and lognormal life distributions, *Technometrics* **18**, 105, (1976).
5. D. L. Maxim, A.D. Hendrickson, and D. E. Cullen, Experimental design for sensitivity testing: the Weibull model, *Technometrics* **19**, 405, (1977).
6. W. Q. Meeker and G. J. Hahn, Asymptotically optimum over-stress tests to estimate the survival probability at a condition with a low expected failure probability, *Technometrics* **19**, 381, (1977).
7. W. B. Nelson and W. Q. Meeker, Theory for optimum accelerated censored life tests for Weibull and extreme value distributions, *Technometrics* **20**, 105, (1978).
8. W. Q. Meeker, A comparison of accelerated life test plans for Weibull and lognormal distributions and type I censoring, *Technometrics* **26**, 157, (1984).
9. W. B. Nelson, *Accelerated Testing: Statistical Models, Test Plans, and Data Analyses*, Wiley, NY, (1990).
10. C. A. Meeter and W. Q. Meeker, Optimum accelerated life tests with a nonconstant sigma, *Technometrics* **36**, 71, (1994).
11. G. B. Yang, Optimum constant-stress accelerated life test plans, *IEEE Transactions on Reliability* **43**, 575, (1994).
12. R. Miller and W. B. Nelson, Optimum simple step-stress plans for accelerated life testing, *IEEE Transactions on Reliability* **32**, 59, (1983).
13. D. S. Bai, M. S. Kim and S. H. Lee, Optimum simple step-stress accelerated life tests with censoring, *IEEE Transactions on Reliability* **38**, 528, (1989).

14. D. S. Bai and Y. R. Chun, Optimum simple step-stress accelerated life tests with competing causes of failure, *IEEE Transactions on Reliability* **40**, 622, (1991).
15. S. W. Chung and D. S. Bai, Optimal designs of simple step-stress accelerated life tests for lognormal lifetime distributions, International Journal of Reliability, *Quality and Safety Engineering* **5**, 315, (1998).
16. I. H. Khamis and J. J. Higgins, Optimum 3-step step-stress tests, *IEEE Transactions on Reliability* **45**, 341, (1996).
17. C. Xiong, Inference on a simple step-stress model with type-II censored exponential data, *IEEE Transactions on Reliability* **47**, 142, (1998).
18. C. Xiong, and G. A. Milliken, Step-stress life-testing with random stress-change times for exponential data, *IEEE Transactions on Reliability* **48**, 141, (1999).
19. C. Xiong and M. Ji, Analysis of grouped and censored data from step-stress life testing, *IEEE Transactions on Reliability* **53**, 22, (2004).
20. J. S. Tomblin, J. D. Tauriello, and S. P. Doyle, A composite material qualification method that results in cost, time, and risk reduction, *Journal of Advanced Materials* **34**, 4, (2002).
21. W. Nelson, Accelerated life testing – step-stress models and data analysis, *IEEE Transactions on Reliability* **R-29(2)**, 103, (1980).
22. H. Hirose, Theoretical foundation for residual lifetime estimation, *Transactions of the Institute of Electrical Engineering of Japan* **116-B(2)**, 168, (1996).
23. Y. Komori, Properties of the Weibull cumulative exposure model, *Journal of Applied Statistics* **33(1)**, 17, (2006).
24. F. Ballani, D. Stonyan, and S. Wolf, On two damage accumulation models and their size effects, *Journal of Applied Statistics* **44**, 99, (2007).
25. W. Q. Meeker and G. H. Hahn. How to plan an accelerated life test-some practical guidelines. Statistical Techniques **10**, ASQC Basic reference in QC, (1985).
26. H. Zhang and E. A. Elsayed, E. A. Nonparametric accelerated life testing based on proportional odds model, *International Journal of Reliability, Quality and Safety Engineering* **13(4)**, 365, (2006).
27. T. Nakamura, Corrected score functions for errors-in-variables models: methodology and application to generalized linear models, *Biometrika* **77**, 127, (1990).
28. G. Y. Yi and J. F. Lawless, A corrected likelihood method for the proportional hazards model with covariates subject to measurement error, *Journal of Statistical Planning and Inference* **137**, 1816, (2007).

OPTIMAL POLICY FOR A TWO-UNIT SYSTEM WITH TWO TYPES OF INSPECTIONS

S. MIZUTANI

Department of Media Informations
Aichi University of Technology
50-2 Manori Nishihazama cho, Gamagori 443-0047, Japan
mizutani@aut.ac.jp

T. NAKAGAWA

Department of Marketing and Information Systems
Aichi Institute of Technology
1247 Yachigusa, Yakusa cho, Toyota, 470-0392, Japan
toshi-nakagawa@aitech.ac.jp

This paper considers an optimal inspection policy for a two-unit system which is checked periodically with two types of inspections: Type-1 inspection has the loss cost which is lower than that of type-2 inspection. However, type-1 inspection is impossible to replace failed unit with a new one. On the other hand, type-2 inspection can replace failed unit with a new one. However, its loss cost is higher than that of type-1 inspection. Therefore, type-1 inspection checks the system so frequently more than type-2 inspection. We consider the time from the beginning of system operation to the replacement of a failed system. Then, the mean time and the expected cost per unit of time are derived analytically. Further, numerical examples are given when the failure time distributions are exponential.

1. Introduction

In recent years, many systems such as digital control devices for information processing have become widely used. Thus, the improvement of their reliability has become necessary and important. Therefore, it is greatly important to check systems and detect their failures by inspection. For the purpose, many researches of optimal inspection policies are proposed in reliability theory areas [1–3].

Several inspection models for a standby unit and a multi-unit system with some types of inspections have been proposed by many authors [4–12]. For example, Ito and Nakagawa considered the inspection models for a storage system which consists of many units [13,14], and the model for a digital control unit of gas-turbine engines which consists of a two-unit system [15].

In this paper, we consider a two-unit system such as digital control devices which consists of a main unit and a spare unit [16], and has input and output sequentially [17] (Figure 1): A main unit is connected with the part of output and is used to control an objective system such as engines of an airplane. A spare unit operates as a standby unit, and has the same function as a main unit. Both units are checked periodically to detect their failures. When we detect a failure of a main unit, a spare unit operates in place of a main unit if it does not fail, and is connected with the part of output. Note that if both units such as engines of an airplane fail then the engine outputs constant signals for fail safe.

Suppose that the periodic inspection is classified with two-types of inspections [18]: Type-1 inspection can detect a failure, although it is impossible to replace the failed unit. Type-2 inspection can detect a failure and can replace the failed unit with a new one. When a failed unit is replaced and maintained by type-2 inspection, the system becomes like new and starts to operate again as a two-unit system. However, the cost for one check of type-2 inspection is higher than that of type-1 inspection. Therefore, type-1 inspection checks the system so frequently more than type-2 inspection.

This paper considers two types of inspections for a two-unit system which is checked periodically. Introducing the loss cost elapsed between a failure and its detection by type-1 or type-2 inspection [16,18], we obtain the expected cost using the inspection theory [1,19,20]. We derive analytically an optimal number of type-1 inspections between type-2 inspections which minimizes the expected cost rate. Numerical examples are finally given when the failure time distributions are exponential.

2. Expected Cost

Consider the periodic inspection policy for a two-unit system with two types of inspections (Figure 1), and make the following assumptions:

(1) A main unit has a general failure distribution $F_1(t)$ with finite mean $1/\lambda_1$, and a spare unit has a general failure distribution $F_2(t)$ with finite mean $1/\lambda_2$, where $1/\lambda_i \equiv \int_0^\infty \overline{F}_i(t)dt$ and $\overline{F}_i(t) \equiv 1 - F_i(t)$ ($i = 1, 2$).

(2) The system is checked at periodic time jT ($j = 1, 2, \dots$) by type-1 inspection and by type-2 inspection at every n times of type-1 inspection, i.e., type-2 inspection is done at time knT ($k = 1, 2, \dots$)

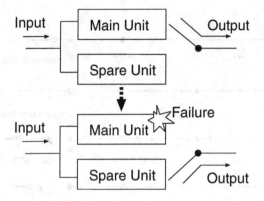

Figure 1. Two-unit system with main unit and spare unit.

(Figure 2). When type-2 inspection is done, type-1 one is not done, and the inspection times are negligible.

(3) When a main or spare unit has failed, the failure is detected by the next type-1 or type-2 inspection. When type-1 inspection detects a failure of a main unit, a spare unit operates in place of a main unit if it does not fail. When type-2 inspection detects failures of both units, the system is replaced and becomes like as new.

(4) Cost c_1 is the cost for one check of type-1 inspection and c_2 is the difference cost between cost for one check of type-1 inspection and that of type-2 inspection. That is, the cost of type-2 inspection is $c_1 + c_2$. Further, c_3 is the loss cost per unit of time for the time elapsed between a failure and its replacement by type-1 or type-2 inspection.

(5) Cost c_4 is the replacement cost for a failed system.

The process from the beginning of system operation to the replacement of a failed system is classified into the following five cases (see Figure 2):
(i) A main unit fails at time t and a spare unit does not fail until the next type-2 inspection (Case (i)).
Then, its probability is

$$\sum_{k=0}^{\infty} \sum_{j=0}^{n-1} \overline{F}_2((k+1)nT) \int_{knT+jT}^{knT+(j+1)T} dF_1(t). \tag{1}$$

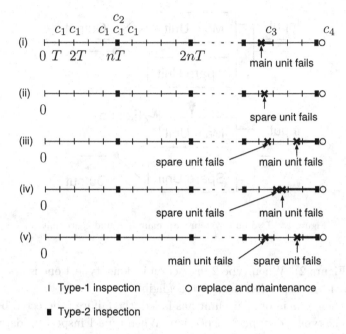

Figure 2. Process from beginning of system operation to the replacement by type-2 inspection.

Thus, the mean time to replacement is

$$\sum_{k=0}^{\infty}\sum_{j=0}^{n-1}(k+1)nT\,\overline{F}_2((k+1)nT)\int_{knT+jT}^{knT+(j+1)T}dF_1(t), \tag{2}$$

and the expected cost to replacement is

$$\sum_{k=0}^{\infty}\sum_{j=0}^{n-1}\overline{F}_2((k+1)nT)$$
$$\times\int_{knT+jT}^{knT+(j+1)T}\{(k+1)nc_1+(k+1)c_2+[knT+(j+1)T-t]c_3\}\,dF_1(t). \tag{3}$$

(ii) A spare unit fails at time t and a main unit does not fail until the next type-2 inspection (Case (ii)).

Then, its probability is

$$\sum_{k=0}^{\infty}\sum_{j=0}^{n-1}\overline{F}_1((k+1)nT)\int_{knT+jT}^{knT+(j+1)T}dF_2(t). \tag{4}$$

Thus, the mean time to replacement is

$$\sum_{k=0}^{\infty}\sum_{j=0}^{n-1}(k+1)nT\,\overline{F}_1((k+1)nT)\int_{knT+jT}^{knT+(j+1)T}dF_2(t), \qquad (5)$$

and the expected cost to replacement by type-2 inspection is

$$\sum_{k=0}^{\infty}\sum_{j=0}^{n-1}\overline{F}_1((k+1)nT)\int_{knT+jT}^{knT+(j+1)T}\{(k+1)nc_1+(k+1)c_2\}\,dF_2(t). \qquad (6)$$

(iii) A main unit fails at time t_1 $(knT+jT < t_1 \le (k+1)nT)$, and a spare unit fails at time t_2 $(knT+jT < t_2 \le knT+(j+1)T)$ (Case (iii) and (iv)). Then, its probability is

$$\sum_{k=0}^{\infty}\sum_{j=0}^{n-1}\int_{knT+jT}^{(k+1)nT}dF_1(t_1)\int_{knT+jT}^{knT+(j+1)T}dF_2(t_2). \qquad (7)$$

Thus, the mean time to replacement is

$$\sum_{k=0}^{\infty}\sum_{j=0}^{n-1}(k+1)nT\int_{knT+jT}^{(k+1)nT}dF_1(t_1)\int_{knT+jT}^{knT+(j+1)T}dF_2(t_2), \qquad (8)$$

and the expected cost to replacement is

$$\sum_{k=0}^{\infty}\sum_{j=0}^{n-1}\int_{knT+jT}^{(k+1)nT}\{(k+1)nc_1+(k+1)c_2+[(k+1)nT-t_1]c_3\}\,dF_1(t_1)$$

$$\times \int_{knT+jT}^{knT+(j+1)T}dF_2(t_2). \qquad (9)$$

(iv) A main unit fails a time t_1 $(knT+jT < t_1 < knT+(j+1)T)$ and a spare unit fails at time t_2 $(knT+(j+1)T < t_2 < (k+1)nT)$ (Case (v)). Then, its probability is

$$\sum_{k=0}^{\infty}\sum_{j=0}^{n-1}\int_{knT+jT}^{knT+(j+1)T}dF_1(t_1)\int_{knT+(j+1)T}^{(k+1)nT}dF_2(t_2). \qquad (10)$$

Thus, the mean time to replacement is

$$\sum_{k=0}^{\infty}\sum_{j=0}^{n-1}(k+1)nT\int_{knT+jT}^{knT+(j+1)T}dF_1(t_1)\int_{knT+(j+1)T}^{(k+1)nT}dF_2(t_2), \qquad (11)$$

and the expected cost to the replacement by type-2 inspection is

$$\sum_{k=0}^{\infty} \sum_{j=0}^{n-1} \int_{knT+jT}^{knT+(j+1)T} \int_{knT+(j+1)T}^{(k+1)nT} \{(k+1)nc_1 + (k+1)c_2$$
$$+ [knT + (j+1)T - t_1 + (k+1)nT - t_2]c_3\}dF_2(t_2)\,dF_1(t_1). \quad (12)$$

Clearly, $(1) + (4) + (7) + (10) = 1$.

Therefore, the mean time to replacement is

$$l(n) = \sum_{k=0}^{\infty} nT\overline{F}_1(knT)\overline{F}_2(knT). \quad (13)$$

Further, the expected cost to replacement is

$$\widetilde{C} = \sum_{k=0}^{\infty} (nc_1 + c_2)\overline{F}_1(knT)\overline{F}_2(knT) + c_3 V(n,T) + c_4, \quad (14)$$

where

$$V(n,T) \equiv T \sum_{k=0}^{\infty} \sum_{j=0}^{n-1} \overline{F}_1(knT + jT)\overline{F}_2((k+1)nT)$$

$$- \sum_{k=0}^{\infty} \overline{F}_2((k+1)nT) \int_{knT}^{(k+1)nT} \overline{F}_1(t)dt$$

$$+ \sum_{k=0}^{\infty} \sum_{j=0}^{n-1} [\overline{F}_2(knT+jT) - \overline{F}_2(knT+(j+1)T)]$$

$$\times \left[(nT-jT)\overline{F}_1(knT+jT) - \int_{knT+jT}^{(k+1)nT} \overline{F}_1(t)dt\right]$$

$$+ \sum_{k=0}^{\infty} \sum_{j=0}^{n-1} [\overline{F}_2(knT + (j+1)T) - \overline{F}_2((k+1)nT)]$$

$$\times \left[T\overline{F}_1(knT+jT) - \int_{knT+jT}^{knT+(j+1)T} \overline{F}_1(t)dt\right]$$

$$+ \sum_{k=0}^{\infty} \sum_{j=0}^{n-1} [\overline{F}_1(knT+jT) - \overline{F}_1(knT+(j+1)T)]$$

$$\times \left\{[nT - (j+1)T]\overline{F}_2(knT+(j+1)T) - \int_{knT+(j+1)T}^{(k+1)nT} \overline{F}_2(t)dt\right\}$$

$$= \sum_{k=0}^{\infty} \left\{ \overline{F}_2(knT) \int_{knT}^{(k+1)nT} [\overline{F}_1(knT) - \overline{F}_1(t)]dt \right.$$
$$\left. + \sum_{j=0}^{n-1} [\overline{F}_1(knT+jT) - \overline{F}_1(knT+(j+1)T)] \int_{knT+(j+1)T}^{(k+1)nT} \overline{F}_2(t)dt \right\}. \quad (15)$$

Therefore, from (13) and (15), the expected rate $C(n)$ is given by

$$C(n) \equiv \frac{\widetilde{C}(n)}{l(n)} = \frac{nc_1 + c_2}{nT} + \frac{c_3 V(n,T) + c_4}{\sum_{k=0}^{\infty} nT\overline{F}_1(knT)\overline{F}_2(knT)}. \quad (16)$$

In particular, $F_1(t) = 1 - e^{-\lambda_1 t}$ and $F_2(t) = 1 - e^{-\lambda_2 t}$, (13) is rewritten by

$$\frac{nT}{1 - e^{-(\lambda_1+\lambda_2)nT}}. \quad (17)$$

and especially, when $c_4 = 0$, $C(n)$ is rewritten as

$$C(n) = \frac{1}{nT}\left[c_2 - c_3\left\{\frac{1-e^{-\lambda_1 nT}}{\lambda_1} - \frac{1}{\lambda_2}\left[e^{-\lambda_2 T}(1-e^{-\lambda_1 T})\frac{1-e^{-(\lambda_1+\lambda_2)nT}}{1-e^{-(\lambda_1+\lambda_2)T}}\right.\right.\right.$$
$$\left.\left.\left. - e^{-\lambda_2 nT}(1-e^{-\lambda_1 nT})\right]\right\}\right] + \frac{c_1}{T} + c_3 \quad (n=1,2,\dots). \quad (18)$$

3. Optimal Policy

We find an optimal number n^* which minimizes $C(n)$ in (18). For this purpose, we consider the optimization problem which minimizes $\widetilde{C}(n)$ given by

$$\widetilde{C}(n) = \frac{1}{nT}\left\{c_2 - c_3\left[\frac{1-e^{-\lambda_1 nT}}{\lambda_1} - \frac{1-e^{-\lambda_2 nT}}{\lambda_2} + \frac{(1-A)}{\lambda_2}(1-e^{-(\lambda_1+\lambda_2)nT})\right]\right\}$$
$$(n=1,2,\dots). \quad (19)$$

where

$$A \equiv \frac{e^{-\lambda_2 T}(1-e^{-\lambda_1 T})}{1-e^{-(\lambda_1+\lambda_2)T}} < 1.$$

Clearly,

$$\widetilde{C}(1) = \frac{1}{T}\left\{c_2 - c_3\left[\frac{1-e^{-\lambda_1 T}}{\lambda_1} - \frac{1-e^{-\lambda_2 T}}{\lambda_2} + \frac{(1-A)}{\lambda_2}(1-e^{-(\lambda_1+\lambda_2)T})\right]\right\},$$

$$\widetilde{C}(\infty) \equiv \lim_{n\to\infty} \widetilde{C}(n) = 0.$$

Forming the inequality $\widetilde{C}(n+1) - \widetilde{C}(n) \geq 0$,

$$\frac{1-e^{-\lambda_1 nT}}{\lambda_1} - \frac{1-e^{-\lambda_2 nT}}{\lambda_2} + \frac{1-e^{-\lambda_2 T}}{\lambda_2(1-e^{-(\lambda_1+\lambda_2)T})}[1-e^{-(\lambda_1+\lambda_2)nT}]$$
$$-n\left\{e^{-\lambda_1 nT}\frac{1-e^{-\lambda_1 T}}{\lambda_1} - \frac{1-e^{-\lambda_2 T}}{\lambda_2}[e^{-\lambda_2 nT} - e^{-(\lambda_1+\lambda_2)nT}]\right\} \geq \frac{c_2}{c_3}. \quad (20)$$

Letting denote the left-hand side of (20) by $L_1(T)$,

$$L_1(1) = (1-e^{-\lambda_1 T})\left[\frac{1-e^{-\lambda_1 T}}{\lambda_1} + \frac{e^{-\lambda_2 T}(1-e^{-\lambda_2 T})}{\lambda_2}\right],$$

$$L_1(\infty) \equiv \lim_{n\to\infty} L_1(n) = \frac{1}{\lambda_1} - \frac{A}{\lambda_2}.$$

Therefore, if $c_3(1/\lambda_1 - A/\lambda_2) > c_2$ then there exists an optimum n^* ($1 \leq n^* < \infty$) which minimizes $\widetilde{C}(n)$.

From $L(n+1) - L(n) \geq 0$,

$$(n+1)\left\{\frac{(1-e^{-\lambda_1 T})^2}{\lambda_1}e^{-\lambda_1 nT} - \frac{(1-e^{-\lambda_2 T})^2}{\lambda_2}e^{-\lambda_2 nT}\right.$$
$$\left.+\frac{1-e^{-\lambda_2 T}}{\lambda_2}(1-e^{-(\lambda_1+\lambda_2)T})e^{-(\lambda_1+\lambda_2)nT}\right\}. \quad (21)$$

When $\lambda_1 = \lambda_2$, it is easily proved that $L(n)$ is a strictly increasing function of n.

Putting $nT = x$, we can rewrite $\widetilde{C}(n)$ as follows:

$$\widetilde{C}(x) = \frac{1}{x}\left\{c_2 - c_3\left[\frac{1-e^{-\lambda_1 x}}{\lambda_1} - \frac{A(1-e^{-(\lambda_1+\lambda_2)x})}{\lambda_2} + \frac{e^{-\lambda_2 x}(1-e^{-\lambda_1 x})}{\lambda_2}\right]\right\}. \quad (22)$$

Clearly,

$$\widetilde{C}(0) \equiv \lim_{x\to 0}\widetilde{C}(x) = \infty, \qquad \widetilde{C}(\infty) \equiv \lim_{x\to\infty}\widetilde{C}(x) = 0.$$

Differentiating $\widetilde{C}(x)$ with respect to x and setting it equal to zero,

$$\frac{1}{\lambda_1}[1-(1+\lambda_1 x)e^{-\lambda_1 x}] - \frac{A}{\lambda_2}\left\{1-[1+(\lambda_1+\lambda_2)x]e^{-(\lambda_1+\lambda_2)x}\right\}$$
$$+\frac{(1+\lambda_2 x)e^{-\lambda_2 x}}{\lambda_2} - \frac{[1+(\lambda_1+\lambda_2)x]}{\lambda_2}e^{-(\lambda_1+\lambda_2)x} = \frac{c_2}{c_3}, \quad (23)$$

i.e.,

$$\frac{1}{\lambda_1}[1-(1+\lambda_1 x)e^{-\lambda_1 x}] - \frac{1}{\lambda_2}[1-(1+\lambda_2 x)e^{-\lambda_2 x}]$$
$$+\frac{1}{\lambda_2}(1-A)\left\{1-[1+(\lambda_1+\lambda_2)x]e^{-(\lambda_1+\lambda_2)x}\right\} = \frac{c_2}{c_3}. \quad (24)$$

Letting denote the left-hand side of (24) by $L_2(T)$,

$$L_2(0) \equiv \lim_{x \to 0} L_2(x) = 0, \qquad L_2(\infty) \equiv \lim_{x \to \infty} L_2(x) = \frac{1}{\lambda_1} - \frac{A}{\lambda_2},$$

$$L'(x) = x \left[\lambda_1 e^{-\lambda_1 x} - \lambda_2 e^{-\lambda_2 x} + \frac{(\lambda_1 + \lambda_2)^2}{\lambda_2} e^{-(\lambda_1 + \lambda_2)x}(1 - A) \right].$$

When $\lambda_1 = \lambda_2$, if $1/\lambda_1 - A/\lambda_2 > c_2/c_3$, $L'(x) > 0$, and hence, there exists a unique x^* that minimizes $\tilde{C}(x)$.

Thus, we derive n^* by computing x^* which satisfies (24) and setting $x^* = n^*T$.

4. Numerical Example

We compute numerically optimal number n^* which minimize the expected cost $C(n)$ in (18) when $F_1(t) = 1 - e^{-\lambda_1 t}$, $F_2(t) = 1 - e^{-\lambda_2 t}$. The costs are normalized to c_1 as unit cost, i.e., they are divided by c_1. Note that c_3/c_1 is not so high because we assume the system can operate using constant signals for fail safe when both units has failed.

Table 1 gives optimal n^* which minimize $C(n)$ for $c_2/c_1 = 10, 20, 30, 40, 50, 60, 70, 80, 90, 100$, $c_3/c_1 = 1, 2, 3, 4, 5, 6, 7, 8, 9, 10$ when $T = 1$, $\lambda_1 = 0.01$ and $\lambda_2 = 0.01$, $c_4/c_1 = 0$. This indicates that optimal n^* increase as c_2/c_1 increase, i.e., it is better to lengthen the interval of type-2 inspection when its cost is large. Further, n^* decrease as the loss cost c_3/c_1 increase. This indicates that Type-2 inspection should be done faster than to replace a failed system when c_3 for a failed interval is large.

Table 1. Optimal n^* which minimize $C(n)$ when $T = 1$, $\lambda_1 = 0.01$ and $\lambda_2 = 0.01$, $c_4/c_1 = 0$.

c_2/c_1	c_3/c_1									
	1	2	3	4	5	6	7	8	9	10
10	4	3	3	2	2	2	2	2	2	1
20	6	4	4	3	3	3	2	2	2	2
30	8	5	4	4	3	3	3	3	3	2
40	9	6	5	4	4	4	3	3	3	3
50	10	7	6	5	4	4	4	4	3	3
60	11	8	6	5	5	4	4	4	4	3
70	12	8	7	6	5	5	4	4	4	4
80	13	9	7	6	6	5	5	4	4	4
90	14	10	8	7	6	5	5	5	4	4
100	14	10	8	7	6	6	5	5	5	4

Table 2 gives optimal n^* which minimize $C(n)$ for $c_2/c_1 = 10, 20, 30, 40, 50, 60, 70, 80, 90, 100$, $\lambda_1 = 0.0100, 0.0050, 0.0010, 0.0005, 0.0001$ when

$T = 1$, $\lambda_1 = 0.001$, $c_3/c_1 = 2$, $c_4/c_1 = 0$. This indicates that optimal n^* increase as λ_2 decrease. This shows that if λ_2 is small, *i.e.*, MTTF of a main unit is large then it is not necessary to perform type-2 inspection so frequently.

Table 2. Optimal n^* which minimize $C(n)$ when $T = 1$, and $\lambda_1 = 0.001$, $c_3/c_1 = 2$, $c_4/c_1 = 0$.

λ_2	c_2/c_1								
	10	20	30	40	50	60	70	80	90	100
0.0100	6	8	10	12	13	15	16	17	18	19
0.0050	5	7	9	11	12	13	14	15	16	17
0.0010	3	4	5	6	7	8	8	9	10	10
0.0005	3	4	5	6	6	7	7	8	8	9
0.0001	2	3	4	5	5	6	6	7	7	7

5. Conclusions

This paper has considered the model that a two-unit system is checked periodically by two types of inspections. Type-1 inspection can detect failures of their units, but it can not replace a failed unit. Type-2 inspection can replace a failed unit with a new one, but its cost is higher than that of type-1 inspection. A two-unit system consists of a main unit and spare unit. When a main unit has failed, if a spare unit does not fail then it works in place of a main unit.

We have obtained the expected cost analytically. Numerical examples are given when the failure time distributions are exponential.

We suggest some modified models as follows: (i) When the failure is detected by type-1 inspection, it is not checked by type-1 inspection between type-2 inspections. (ii) When each of units has failed, a two-unit system is replaced as soon as possible. (iii) The failure is classified with two-types of failures. Type-1 failure is detected by type-1 or type-2 inspection, and type-2 failure is detected only by type-2 inspection.

The studies of reliability and inspection for high reliable systems become more important subjects as systems become more complex and large. These formulation and results would be applied to other real systems such as digital control device or digital circuits by suitable modifications.

References

1. R.E. Barlow and F. Proschan, *Mathematical Theory of Reliability*, John Wiley & Son, New York, (1965).

2. N. Kaio, S. Osaki, Optimal inspection policies: A review and comparison, *Journal Mathematics and Analytical Application*, **119**, pp. 3–20, (1986).
3. T. Nakagawa, *Maintenance Theory of Reliability*, Springer-Verlag, London, (2005).
4. T. Nakagawa, Optimum inspection policies for a standby unit, *Journal Operations Research Society Japan*, **23**, pp. 13–26, (1980).
5. N. Kaio, S. Osaki, Optimal inspection policy with two types of imperfect inspection probabilities, *Microelectron Reliability*, **26**, pp. 935–942, (1986).
6. L. C. Thomas, P. A. Jacobs, D. P. Gaverd, Optimal inspection policies for standby systems, *Communication in Statistics: Stochastic Models*, **3**, pp. 259–273, (1987).
7. S. H. Sim, Reliability of standby equipment with periodic testing. *IEEE Transactions on Reliability*, **R-36**, pp. 117–123, (1987).
8. D. Zuckerman, Optimal inspection policy for a multi-unit machine, *Journal of Applied Probability*, **26**, pp. 543–551, (1989).
9. T. Teramoto, T. Nakagawa, M. Motoori, Optimal inspection policy for a parallel redundant system, *Microelectron Reliability*, **30**, pp. 151–155, (1990).
10. N. Kaio, T. Dohi, S. Osaki, Inspection policy with failure due to inspection, *Microelectron Reliability*, **34**, pp. 599–602, (1994).
11. G. Parmigiani, Inspection times for stand-by units, *Journal of Applied Probability*, **31**, pp. 1015–1025, (1994).
12. J. K. Vaurio, Unavailability analysis of periodically tested standby components, *IEEE Transactions on Reliability*, **44**, pp. 512–517, (1995).
13. K. Ito, T. Nakagawa, An optimal inspection policy for a storage system with finite number of inspections, *Journal of Reliability Engineering Association of Japan*, **19**, pp. 390–396, (1997).
14. K. Ito, T. Nakagawa, Optimal inspection policies for a storage system with degradation at periodic tests, *Mathematics and Computer Model* **31**, pp. 191–195, (2000).
15. K. Ito, T. Nakagawa, Optimal self-diagnosis policy for dual redundant FADEC of gas turbine engines, In *Proceedings of ASSM2000*, March 29-30, pp. 83–87, (2000).
16. S. Mizutani, T. Nakagawa, K. Yasui, T. Nishimaki, Comparison-checking diagnosis schemes for a two-unit system, In *The Fourth Asia-Pacific Conference on Industrial Engineering and Management Systems*, pp. A–59, (2002).
17. P. K. Lala, *Self-Checking and Fault Tolerant Digital Design*, Morgan Kaufman Pub., San Francisco, (2001).
18. S. Mizutani, T. Nakagawa, K. Ito, Optimal inspection policies for a self-diagnosis system with two types of inspections, In *Reliability Modeling, Analysis and Optimization*, (Edited by H. Pham), pp. 417–428, World Scientific, New Jersey, (2006).
19. S. Osaki, *Applied Stochastic System Modeling*, Springer Verlag, Berlin, (1992).
20. S. M. Ross, *Applied Probability Models with Optimization Applications*, San Francisco, Holden-Day, (1970).

REDUNDANCY OPTIMIZATION IN MULTI-LEVEL SYSTEM USING META-HEURISTICS

IL HAN CHUNG

Technical Research Institute, Hyundai Rotem Company, 462-18, Sam-Dong, Uiwang-Shi, Gyunggi-Do, 449-910, Korea

WON YOUNG YUN

Department of Industrial Engineering, Pusan National University, 30 Changjeon-Dong, Kumjeong-Ku, Busan 609-735, Korea
wonyan@pusan.ac.kr

HO GYUN KIM

Department of Industrial and Management Engineering, Dong-Eui University, Busanjin-Ku, Busan, 614-714, Korea

Single-level systems have been considered in redundancy allocation problems. Traditionally, we assume typical structures, for example, series, k-out-of-n, series-parallel, and determine how many redundant units should be assigned to each unit in the system structure. In regard to system reliability, it is most effective to duplicate the lowest level objects, because parallel-series systems are more reliable than series-parallel systems. However, it may not be the most cost effective way. In this paper, redundancy is considered at all levels in a series system with multi-levels, and a mixed integer programming model is formulated. Some meta-heuristics (genetic algorithm, simulated annealing, and ant colony algorithm) are considered to solve the problem and some examples are studied.

1. Introduction

Reliability is considered to be one of most important design measures in various systems. Redundancy allocation has been used mainly to enhance system reliability. The redundancy allocation problem, which involves selecting redundancy levels at each subsystem in order to maximize system reliability under several resource constraints, is a well-known combinatorial optimization problem. The problem arises frequently in system designs such as semi-conductor integrated circuits, nanotechnology, and most electronic systems. Several researchers have previously solved redundancy allocation problems, which have several system structures such as series, series-parallel, network, k-out-of-n, and so on and used various optimization approaches and different problem formulations. Dynamic programming [1], Lagrange multiplier [6], heuristic approaches [5, 7, 8, 9, 10, 17], ant colony optimization (ACO) [14, 4], simulated annealing (SA) [16] and genetic algorithm (GA) [2, 3, 20, 21] have been used for the problems.

For a list of recent papers, see a review paper by Kuo and Prasad [18]. In most papers, the objects of redundancy allocation have been limited to single levels because of the well-known design principle that redundancy at the component level is more effective than redundancy at the system level. This is true under some specific assumptions, but Boland and EL-Neweihi [15] showed that this is not true for redundancy problems with non-identical spare parts.

The purpose of this paper is to maximize system reliability by allocating redundancy units under constraints on cost, volume, weight and other variables. Modular redundancy with identical spare parts is addressed by this paper. Modular redundancy can be more effective than component redundancy because in modular systems, duplicating a module composed of several components can be easier, and requires less time and skill, than duplicating each component. Therefore the lower the level of the redundant item and the more spare parts added, the higher the cost of redundancy. Yun et al. [19], Yun and Kim [20], and Yun et al. [21] considered similar problems and proposed several methods. In this paper, we consider the same model as Yun and Kim [20], summarizes some meta-heuristics for solving the mathematical problem, and compare different methods by numerical examples. For assumptions and mathematical model, refer to Yun et al. [20]. In this paper, we explain the redundancy problem briefly.

The system illustrated in Fig. 1 is an example of a system structure with three levels and seven components. The system has three subsystems (A, B, C). Subsystem A and B have two components and subsystem C has a module and a component. Most of the papers related to redundancy optimization problems consider only the lowest-level components to be candidates for redundancy (The left structure of Fig. 1 shows the lowest-level units as candidates for redundancy in the existing papers). However in this paper, the module, subsystems, and system itself are all candidates for redundancy (In the right structure of Fig. 1, all units are candidates for redundancy).

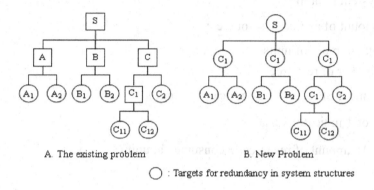

A. The existing problem   B. New Problem

○ : Targets for redundancy in system structures

Figure 1. System hierarchical structures in the existing problems and the new problem.

Before further discussion of the proposed model further, some concepts are defined. The term 'unit' is used as a common name for a system, subsystem, module, and component. For other terms (path, ancestor, offspring, parent, child, sibling, and cousin) refer to Yun et al. [19, 20].

This paper is organized as follows. In Section 2, we explain the concept of the presented model. The developed meta-heuristic method, ACO, is illustrated in Section 3. The results from several numerical experiments are discussed in Section 4. We conclude in Section 5 with some suggestions for further studies.

2. Model Description

This paper considers series systems and parallel redundancy, and assumes failures to be statistically independent. We assume that only a level can be selected for redundancy in a path set. Under these assumptions, the problem of allocating modular redundancy is formulated as follows:

Notation

j_f : a set of ancestor units of unit j

R_s : system reliability

b_r : amount of available resource r

N : total number of units

n_r : number of resources

x_j : number of redundancy allocated to unit j

y_j : 0 or 1 indicator variable

$g_{rj}(x_j)$: amount of resources r consumed at unit j.

$$\text{Max } R_s = \prod_{i=1}^{n}\left(1-y_i(1-R_i)^{x_i}\right) \tag{1}$$

$$\text{s.t. } \sum_{i=1}^{n} y_i g_{ri}(x_i) \leq b_r, \quad r=1,2,\cdots,n_r \tag{2}$$

$$y_j + \sum_{k\in\{j_f\}} y_k = 1, \tag{3}$$

$$y_i = 0 \text{ or } 1, \quad \forall i \tag{4}$$

All $x \geq 1$ and integer j denote the components in the lowest level

The objective is to maximize system reliability. Two prime variables, x_j and y_j, are used. x_j denote the number assigned to unit j. y_j indicate whether unit j is actually used or not. Consequently, $x_j \times y_j$ yields the number of unit j used in the system. Eq. 2 represents the constraints of available resources, such as cost, weight and volume, where all $g_{rj}(x_j)$ are assumed to be linear with the exception of cost. $C(x) = cx + \lambda^x$ is used as a cost function. Eq. 3 is a constraint that only a single unit can be used in a direct line.

In this paper, we compare three meta-heuristics (GA, SA, and ACO) in the proposed redundancy optimization problem. Yun et al. [19] and Yun and Kim [20] already proposed GA and SA algorithms for the same redundancy optimization problem. Thus, in this paper, we suggest an ACO and compare three meta-heuristics using several numerical examples.

3. Ant Colony Optimization

Ant colony optimization (ACO), introduced by Dorigo [11, 12], is a probabilistic technique for solving computational problems which can be simplified to find good paths through graphs. It is inspired by the behavior of ants in finding paths from the colony to food.

In a real world, ants (initially) wander randomly, and upon finding food return to their colony while leaving pheromone trails. If other ants find such a path, they are likely not to keep traveling at random, but to instead follow the trail, return and reinforce it if they eventually find food. Over time, however, the pheromone trail starts to evaporate, thus reducing its indicative strength. The more time it takes for an ant to travel down the path and back again, the more time the pheromones have to evaporate. A short path, in comparison, gets marched over faster, and thus the pheromone density remains high as it is laid on the path as fast as it can evaporate. Pheromone evaporation also has the advantage of avoiding the convergence to a locally optimal solution [22].

To apply ACO to various combinatorial optimization problems, solution representation, state transition rule, and pheromone trail should be determined.

3.1. *Solution representation*

To solve the redundancy optimization problem using ACO, we should build a path corresponding to a solution. In this paper, we use the existing method (Nahas et al. [14]) and define the node and edge.

Three sets of nodes are
- The first set of nodes (N_1) represents the components.

- The second set of nodes (N_2) represents y_j.
- The third set of nodes (N_3) represents x_j.

Two sets of edges are
- The first set of edges is used to connect each component node in the set N_1 to the corresponding nodes in N_2.
- The second set of edges is used to connect the nodes in N_2 to the nodes in N_3.

If the node of N_2, y_j is equal to 0, the node of N_3, x_j is also to be 0. Otherwise ($y_j = 1$), x_j have a value between 1 and P_{max}. Using the nodes and edges defined previously in ACO, we find a solution that is a set of (y_j, x_j) for N basic items.

3.2. State transition rule

An ant for component i chooses y that indicates whether component i is actually used or not, and x denotes the number allocated to component i. When the value of y is 0, x is always 0. Otherwise, when the value of y is 1, x can take any value between 1 and P_{max}.

$$P_{iyx} = \frac{(\tau_{iyx})^\alpha}{\sum_{k=1}^{P_{max}}(\tau_{i1k})^\alpha + (\tau_{i00})^\alpha},$$

$i = 0,1,...,n$, $y = 0,1$, and $x = 1,2,...,P_{max}$, where α is a parameter that controls the weight of the pheromone (τ_{ijk}). P_{i00} is 0.5 and P_{i1x} is $0.5/P_{max}$ in the initial transition probability. If the value of pheromone is updated, the transition probability is changed.

3.3. Method for generating a feasible neighborhood solution

There are some restrictions on the equations 2 and 3 for paths during searching solution using ACO. In this paper, we generate the nodes and edges satisfying the equation 3 but we give a large penalty to pheromone for the solution that does not satisfy the equation 2. The steps to find the solution that satisfies the equation 3 is as follows;

- Step 1: Select one unit randomly.
- Step 2: For the selected unit, P_{i00} are assigned as the value of Y. P_{i1x} is assigned as the probability of X and go to step 3. Otherwise (Y = 0), go to step 4.
- Step 3: For the selected unit, the values of Y and X of ancestor and offspring are set to 0.
- Step 4: If there are some units without assigned value, go to step 1. Otherwise stop this procedure.

The initial values of P_{i00} and P_{i1x} will be changed to improve the objective function while the ants search the solutions.

3.4. Pheromone trail

After all ants go through the procedure, the value of pheromone is updated. The updated pheromone affects the transition probability and guides the ants to find better solutions in next iteration. We update the value of pheromone by the method proposed by Nahas [14] and Zhao [4].

LC is the current iteration count, and CQ is the quantity of ants in each ant colony.

$$\tau^{LC}_{ijk} = \rho \tau^{LC-1}_{ijk} + (1-\rho) \sum_{p=0}^{CQ} \Delta \tau_{ijk}^{p},$$

where ρ is the persistence of the pheromone trail, so that $1-\rho$ represents the evaporation of pheromone from the previous iteration.

$\Delta \tau^{p}_{ijk}$ is the quantity of pheromone with $y_i = j$ and $x_i = k$ of an ant p for component i and are given by

$$\Delta \tau^{p}_{ijk} = Q \cdot Penalty_p \cdot R_s^p,$$

where Q is a positive number and R_s^p is the system reliability for ant p. In general, there are some infeasible solutions and penalties for the infeasible solutions are used to update pheromone as follows:

$$Penalty_p = \prod_{i=0}^{n_r} Min\left[\left(\frac{\sum_{i=1}^n y_i g_{ri}(x_i) - b_r}{b_r}\right)^{\beta}, 1\right]$$

4. Numerical Examples

To compare the proposed methods, we use the examples from Yun and Kim [20]. The cost function is $C(x) = cx + \lambda^x$.

We consider three numerical examples: First, we consider the enumeration method and three heuristics for the redundancy problem in a system structure with three levels and compare the performance of the heuristics. Second, the three heuristics are compared in system structures with four and five levels because it is very difficult to obtain the optimal solutions by the enumeration method. Finally, we conduct a sensitivity analysis using a SA that gives the best solutions.

4.1. Three level case

Table 1 gives the values of system input variables and Table 2 shows the total costs and system reliabilities. The enumeration method gives the optimal solutions. Among three heuristics, the SA gives the best solutions relatively in many cases.

4.2. Four and five level cases

For more comparisons, we consider more complex structures, and Tables 3 and 4 show the values of input variables. Tables 5 and 6 represent the results and the SA gives the best solutions relatively in many cases. But the GA is best in computation time. In GA, it takes long time to satisfy the constraint (2) but in ACO, it takes long time to find the neighborhood solutions satisfying the constraint (3).

4.3. Sensitivity analysis in the SA

As a simple sensitivity analysis to study the effect of input variables, we change the values of additive cost λ, the price of module, and reliabilities of the lowest components. Table 7 shows the cases for sensitivity analysis and Tables 8–11 give the results as follows: the additive cost for module 13 does not affect the optimal solutions but the cost of component 131 affects the optimal solutions. If the price of module 13 is high, it is better to consider redundancy at the lower level. Finally, an increase in reliability of component 131 gives the higher system reliability.

5. Conclusion

This paper dealt with a redundancy optimization problem in which all units at multi-levels can be candidates for redundancy optimization in series systems. We considered three heuristics (SA, GA and ACO) in which two heuristics have been proposed in the previous research and compared their performance to find the optimal solution. With numerical examples, we found that the heuristics give reasonable solutions and the SA generates the best solutions relatively in many cases. But we should remember that the performance depends strongly on the used model parameters in algorithms.

For the further research, we will consider various redundancy problems, apply the heuristics and compare their performance. Even if we consider the static redundancy problems, the redundancy problems in dynamic reliability cases may be a promising area and in that case, operation problem can also be considered together with redundancy problem (design of system structure).

Table 1. The input data for 3 level structure (Rel.: Reliability).

Unit	Parent unit	Rel.	Price	Additive cost parameter
1(system)	-	0.40029	72	2
11	1	0.72675	26	2
12	1	0.76500	19	3
13	1	0.72000	21	2
111	11	0.90000	5	3
112	11	0.95000	6	4
113	11	0.85000	5	4
121	12	0.90000	6	4
122	12	0.85000	7	4
131	13	0.90000	8	3
132	13	0.80000	7	4

Table 2. Cost limit, total cost, and system reliability of the optimal solution for 3 level structure (*: the same result with the optimal solution).

Cost of resource	Optimal		GA		SA		AOC	
	Total Cost	Rel.	Total Cost	Rel.	Total Cost	Rel.	Total Cost	Rel.
150	149	0.8057	149	0.8053	149	*0.8057	140	0.7178
160	159	0.8476	149	0.8053	158	0.8309	158	0.8309
170	169	0.8511	168	0.8455	169	*0.8511	168	0.8455
180	175	0.8668	174	0.8547	175	*0.8668	177	0.8609
190	184	0.8995	186	0.8878	186	0.8878	185	0.8758
200	199	0.9010	186	0.8878	199	*0.9010	199	*0.9010
210	210	0.9077	204	0.9050	202	0.9136	204	0.9052
220	215	0.9272	204	0.9050	215	*0.9272	215	*0.9272
230	228	0.9319	204	0.9050	228	*0.9319	215	0.9272
240	228	0.9319	233	0.9197	228	*0.9319	228	*0.9319
250	241	0.9457	241	*0.9457	241	*0.9457	231	0.9284
260	257	0.9469	241	0.9457	241	0.9457	241	0.9457
270	270	0.9609	270	*0.9609	270	*0.9609	241	0.9457
280	270	0.9609	271	0.9608	270	*0.9609	275	0.9600
290	270	0.9609	271	0.9608	270	*0.9609	270	*0.9609
300	270	0.9609	271	0.9608	270	*0.9609	280	0.9555
310	304	0.9754	304	*0.9754	304	*0.9754	299	0.9606
320	304	0.9754	304	*0.9754	304	*0.9754	304	*0.9754
330	304	0.9754	304	*0.9754	304	*0.9754	304	*0.9754
340	304	0.9754	304	*0.9754	304	*0.9754	333	0.9742

Table 3. The input data for 4 level structure.

Unit	Parent unit	Rel.	Price	Additive cost parameter
1(System)	-	0.3769	120	2
11	1	0.6441	38	2
12	1	0.5852	47	2
111	11	0.7695	18	2
112	11	0.8370	15	3
121	12	0.8835	13	2
122	12	0.7914	17	2
123	12	0.8370	12	2
1111	111	0.9500	5	4
1112	111	0.9000	6	4
1113	111	0.9000	4	3
1121	112	0.9300	7	3
1122	112	0.9000	5	3
1211	121	0.9300	4	2
1212	121	0.9500	5	3
1221	122	0.8500	4	3
1222	122	0.9800	5	3
1223	122	0.9500	6	4
1231	123	0.9300	4	3
1232	123	0.9000	5	3

Table 4. Cost limit, total cost, and system reliability of the optimal solution for 4 level structure (* : The best result among three heuristics).

Cost of resource	GA		SA		AOC	
	Total Cost	Rel.	Total Cost	Rel.	Total Cost	Rel.
160	165	0.8113	158	*0.8236	160	0.7389
180	179	0.8621	169	*0.8713	180	0.8151
200	200	0.8671	189	*0.9136	200	0.8931
220	222	0.9317	210	*0.9363	219	0.9206
240	238	*0.9530	236	0.9452	229	0.9370
260	259	0.9341	257	*0.9682	255	0.9460
280	279	*0.9678	279	*0.9678	276	0.9676
300	299	0.9699	300	0.9736	295	*0.9753
320	310	0.9780	308	0.9813	320	*0.9823
340	329	0.9794	333	*0.9884	328	0.9810
360	360	0.9796	354	*0.9898	348	0.9872
380	373	*0.9917	376	0.9901	376	0.9901
400	388	0.9902	397	*0.9927	397	0.9865
420	415	0.9916	415	*0.9932	412	0.9862
440	438	*0.9928	438	*0.9928	430	0.9909
460	456	0.9958	454	*0.9962	460	0.9947
480	467	*0.9968	478	0.9952	479	0.9959
500	495	*0.9974	479	0.9959	473	0.9956

Table 5. The input data for 5 level structure.

Unit	Parent unit	Rel.	Price	Additive cost parameter
1(System)		0.1473	150	2
11	1	0.4840	62	2
12	1	0.3044	78	2
111	11	0.6403	28	2
112	11	0.7559	28	2
121	12	0.5560	35	2
122	12	0.5475	40	2
1111	111	0.7650	12	2
1112	111	0.8370	14	3
1121	112	0.8370	13	3
1122	112	0.9031	13	2
1211	121	0.7268	20	2
1212	121	0.7650	13	2
1221	122	0.8835	11	2
1222	122	0.8100	14	3
1223	122	0.7650	12	2
11111	1111	0.9000	5	3
11112	1111	0.8500	4	3
11121	1112	0.9300	6	3
11122	1112	0.9000	5	3
11211	1121	0.9000	5	3
11212	1121	0.9300	5	3
11221	1122	0.9800	6	4
11222	1122	0.9500	5	3
11223	1122	0.9700	6	3
12111	1211	0.9500	5	3
12112	1211	0.8500	4	3
12113	1211	0.9000	6	4
12121	1212	0.9000	5	3
12122	1212	0.8500	5	3
12211	1221	0.9500	5	4
12212	1221	0.9300	4	3
12221	1222	0.9000	6	3
12222	1222	0.9000	5	3
12231	1223	0.9000	6	3
12232	1223	0.8500	4	3

Table 6. Cost limit, total cost, and system reliability of the optimal solution for 5 level structure (* : The best result among three heuristics).

Cost of resource	GA		SA		AOC	
	Total Cost	Rel.	Total Cost	Rel.	Total Cost	Rel.
220	220	0.3514	220	*0.3945	219	0.3408
240	239	0.3893	240	*0.4588	240	0.4279
260	260	*0.5460	258	0.4904	250	0.4342
280	279	0.5337	270	*0.5759	272	0.5124
300	298	*0.7082	295	0.6166	289	0.5455
320	320	0.7578	318	0.7018	301	0.5902
340	339	0.7832	339	*0.7950	329	0.6613
360	360	*0.8416	360	*0.8416	336	0.6480
380	379	0.8364	377	*0.8755	380	0.7263
400	397	0.8843	392	*0.8862	397	0.7878
420	420	0.8415	420	*0.8443	410	0.8151
440	436	*0.8835	438	0.8293	440	0.8202
460	458	0.8718	457	*0.8841	449	0.8415
480	480	0.9011	469	*0.9030	476	0.8609
500	498	0.9253	499	*0.9308	482	0.8913
520	514	0.9305	519	*0.9399	495	0.8975
540	538	0.9452	538	*0.9494	529	0.9025
560	554	0.9427	556	*0.9575	552	0.9328
580	578	0.9488	579	*0.9530	544	0.9255
600	599	0.9626	567	0.9390	600	0.9363

Table 7. The input data for sensitivity analysis.

	Price of Module 13	Additive Cost of Module 13	Additive Cost of Component 131	Reliability of Component 131
Case 1	25	1	1	0.80
Case 2	23	2	2	0.85
Case 3	21	3	3	0.90
Case 4	19	4	4	0.95
Case 5	17	5	5	0.99

Table 8. Results for different additive costs of module 13.

Additive Cost	System Rel.	Total Cost	y/x	1	11	12	13	111	112	113	121	122	131	132
1	0.8309	158	y	0	1	1	0	0	0	0	0	0	1	1
			x	0	2	2	0	0	0	0	0	0	2	2
2	0.8309	158	y	0	1	1	0	0	0	0	0	0	1	1
			x	0	2	2	0	0	0	0	0	0	2	2
3	0.8309	158	y	0	1	1	0	0	0	0	0	0	1	1
			x	0	2	2	0	0	0	0	0	0	2	2
4	0.8309	158	y	0	1	1	0	0	0	0	0	0	1	1
			x	0	2	2	0	0	0	0	0	0	2	2
5	0.8309	158	y	0	1	1	0	0	0	0	0	0	1	1
			x	0	2	2	0	0	0	0	0	0	2	2

Table 9. Results for different additive costs of component 131.

Additive Cost	System Rel.	Total Cost	y/x	1	11	12	13	111	112	113	121	122	131	132
1	0.8384	158	y	0	1	1	0	0	0	0	0	0	1	1
			x	0	2	2	0	0	0	0	0	0	3	2
2	0.8309	153	y	0	1	1	0	0	0	0	0	0	1	1
			x	0	2	2	0	0	0	0	0	0	2	2
3	0.8309	158	y	0	1	1	0	0	0	0	0	0	1	1
			x	0	2	2	0	0	0	0	0	0	2	2
4	0.8253	160	y	0	1	0	1	0	0	0	1	1	0	0
			x	0	2	0	2	0	0	0	2	2	0	0
5	0.8253	160	y	0	1	0	1	0	0	0	1	1	0	0
			x	0	2	0	2	0	0	0	2	2	0	0

Table 10. Results for different prices of module 13.

Price	System Rel.	Total Cost	y/x	1	11	12	13	111	112	113	121	122	131	132
17	0.8405	158	y	0	0	1	1	1	1	1	0	0	0	0
			x	0	0	2	2	2	2	2	0	0	0	0
19	0.8309	158	y	0	1	1	0	0	0	0	0	0	1	1
			x	0	2	2	0	0	0	0	0	0	2	2
21	0.8309	158	y	0	1	1	0	0	0	0	0	0	1	1
			x	0	2	2	0	0	0	0	0	0	2	2
23	0.8309	158	y	0	1	1	0	0	0	0	0	0	1	1
			x	0	2	2	0	0	0	0	0	0	2	2
25	0.8309	158	y	0	1	1	0	0	0	0	0	0	1	1
			x	0	2	2	0	0	0	0	0	0	2	2

Table 11. Results for different reliabilities of component 131.

Rel.	System Rel.	Total Cost	y/x	1	11	12	13	111	112	113	121	122	131	132
0.80	0.8057	158	y	0	1	1	0	0	0	0	0	0	1	1
			x	0	2	2	0	0	0	0	0	0	2	2
0.85	0.8204	158	y	0	1	1	0	0	0	0	0	0	1	1
			x	0	2	2	0	0	0	0	0	0	2	2
0.90	0.8309	158	y	0	1	1	0	0	0	0	0	0	1	1
			x	0	2	2	0	0	0	0	0	0	2	2
0.95	0.8439	160	y	0	1	0	1	0	0	0	1	1	0	0
			x	0	2	0	2	0	0	0	2	2	0	0
0.99	0.8567	160	y	0	1	0	1	0	0	0	1	1	0	0
			x	0	2	0	2	0	0	0	2	2	0	0

Acknowledgements This work was supported by "Research Center for Logistics Information Technology (LIT)" hosted by the Ministry of Education & Human Resources Development in Korea.

References

1. C.F. Woodhouse, Optimal redundancy allocation by dynamic programming, *IEEE Transactions on Reliability*, 21(1), 60–62(1972).

2. D.W. Coit and A.E. Smith, Reliability optimization of series-parallel systems using a genetic algorithm, *IEEE Transactions on Reliability*, 45(2), 254–266(1996).
3. D.W. Coit and A.E. Smith, Redundancy allocation to maximize a lower percentile of the system time to failure distribution, *IEEE Transactions on Reliability*, 47(1), 79–87(1998).
4. J.H. Zhao, Z. Liu, and M.T. Dao, Reliability optimization using multi objective ant colony system approaches, *Reliability Engineering and System Safety*, 92, 109–120(2007).
5. J. Li, A bound heuristic algorithm for solving reliability redundancy optimization, *Microelectronics and Reliability*, 36(3), 335–339(1996).
6. K.B. Misra, Reliability optimization of a series-parallel system, *IEEE Transactions on Reliability*, 21, 230–238(1972).
7. K.B. Misra, A simple approach for constrained redundancy optimization problems, *IEEE Transactions on Reliability*, 21(1), 30–34(1972).
8. K. Gopal, K.K. Aggarwal, and J.S. Gupta, An improved algorithm for reliability optimization, *IEEE Transactions on Reliability*, 27(5), 325–328(1978).
9. K.K. Aggawal, Redundancy optimization in general systems, *IEEE Transactions on Reliability*, 25(5), 330–332(1976).
10. K.K. Aggawal, J.S. Gupta, and K.B. Misra, A new heuristic criterion for solving a redundancy optimization, *IEEE Transactions on Reliability*, 24, 86–87(1975).
11. M. Dorigo, Optimization learning and natural algorithms, Doctoral Dissertation, Politecnico di Milano(1992).
12. M. Dorigo, V. Maniezzo, and A. Colorni, Ant System: Optimization by a Colony of Cooperating Agents, *IEEE Transactions on Systems, Man, and Cybernetics–Part B*, 26 (1), 29–41(1996).
13. M. Gen, and R. Cheng, Genetic algorithms and engineering design, John Wiley and Sons(1996).
14. N. Nahas, M. Nourelfath, and D. Ait-Kadi, Coupling ant colony and the degraded ceiling algorithm for the redundancy allocation problem of series-parallel system, *Reliability Engineering and System Safety*, 92, 211–222(2007).
15. P. Boland and E. EL-Neweihi, Component redundancy vs. system redundancy in the hazard rate ordering, *IEEE Transactions on Reliability*, 44(4), 614–619(1995).
16. V. Ravi, B.S.N. Muty, and P.J. Reddy, Non-equilibrium simulated annealing algorithm applied to reliability optimization of complex systems, *IEEE Transactions on Reliability*, 46, 233–239(1997).
17. W. Kuo, C.L. Hwang, and F.A. Tillman, A note on heuristic methods in optimal system reliability, *IEEE Transactions on Reliability*, 27(5), 320–324(1978).
18. W. Kuo and V.R. Prasad, An annotated overview of system-reliability optimization, *IEEE Transactions on Reliability*, 49(2), 176–187(2000).
19. W.Y. Yun, I.H. Chung, and H.G. Kim, Redundancy optimization in multi-level system with SA algorithm, *Proceedings of the 2nd Asian International Workshop*, 185–192 (2006).

20. W.Y. Yun and J.W. Kim, Multi-level redundancy optimization in series systems, *Computers & Industrial Engineering*, 46, 337–346(2004).
21. W.Y. Yun, Y.M. Song, and H.G. Kim, Multiple multi-level redundancy allocation in series systems, *Reliability Engineering and System Safety*, 92, 308–313(2007).
22. http://en.wikipedia.org/wiki/Ant_colony_algorithm.

OPTIMAL CENSORING POLICIES FOR THE OPERATION OF A DAMAGE SYSTEM

KODO ITO

Institute of Consumer Sciences and Human Life,
Kinjo Gakuin University,
1723 Omori 2-chome, Moriyama-ku,
Nagoya, Aichi, 463-8521, Japan
itokodo@gmail.com

TOSHIO NAKAGAWA

Department of Marketing and Information Systems,
Aichi Institute of Technology,
1247 Yachigusa, Yakusa-cho,
Toyota, Aichi, 470-0392, Japan,
toshi-nakagawa@aitech.ac.jp

Aged fossil-fired power plants are on the great increase in Japan and they need the maintenance for their steady operations. The preventive maintenance (PM) of such systems is requisite to avert the grave trouble such as the emergency stop of power supply. The operating condition of systems after PM cannot return to brand-new because the cumulative fatigue damage of system parts remains and their damage increases sharply after PM. Such degradation of systems has to be evaluated when the maintenance plan is established. In this paper, a system is operated continuously and the PM is done at a prespecified schedule when the cumulative damage level is below a managerial one. When the total damage level exceeds a certain critical one, a system fails. The expected profit until a certain PM time is obtained, and the optimal operation censoring policy is derived, using the theory of cumulative processes.

1. Introduction

A number of aged fossil-fired power plants are growing in Japan. For example, 33% of these systems are currently operated from 150,000 (17 years) to 199,999 hours (23 years), and 26% of them are above 200,000 hours [1]. Although Japanese government loosened legal controls of the electric power industry, most companies hesitate to build new systems and tend to operate existing systems efficiently. Such situation is caused by the advance of microscopic inspection technologies. Today, the residual life of essential parts of power plants such as boiler tubes and turbine blades are evaluated precisely and efficiently [2,3].

The well-thought-out maintenance plans are indispensable to operate these aged systems without any serious troubles such as the emergency stop of operation. The importance of maintenance for aged systems is much higher than that for new ones, because occurrence probabilities of severe troubles increase and new failure phenomena might appear according to the degradation of systems. Furthermore, the system operation should halt at adequate period times, considering both the profit of operation and the loss of scheduled maintenance and unscheduled failure.

The maintenance is classified into the preventive maintenance (PM) and the corrective maintenance (CM). Many authors have studied PM policies for systems because the CM cost at failure is much higher than the PM one and the consideration of effective PM is significant [4]. The occurrences of failures have been discussed by utilizing the cumulative process [5,6]. Some aspects of damage models from reliability viewpoints were discussed [7]. The PM policies where a system is replaced before failure at time T, at shock N, and at damage K were considered [8-13].

For establishing the cost-effective PM, the precise life estimation analysis is necessary. A cumulative damage model for system life under varying stress in the accelerated test or in the actual field use was considered [14,15]. Lots of statistical methods of accelerated test to establish reliability parameters of system were assessed [16].

Most systems consist of a wide variety of mechanical parts such as power boiler tube, compressor, combustor, steam and gas turbine blade. Some parts suffer high temperature at operation and thermal damages are accumulated in these parts. PM is performed periodically before these damages cause serious failures. The condition after PM cannot return to the brand-new condition because some cumulative fatigue damage of parts remains after PM [17]. In past PM studies and cumulative damage models, the condition after PM is supposed to be brand-new. In the actual system maintenance, the remaining damage after PM should be considered.

In this paper, the PM is done at a prespecified schedule when the total damage is below a managerial level and some damage remains after the PM. When the cumulative damage level is between a managerial level and a certain critical one, the system stops its operation and is overhauled. When the cumulative damage level exceeds a certain critical one, the system fails

and the critical level becomes lower at every PM. The expected profit is
considered and an optimal operation censoring policy is derived.

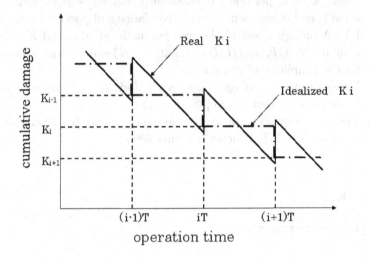

Figure 1. Schematic diagram of real K_i and idealized K_i.

2. Model 1

We consider the following PM policy (see Figures 1 and 2) for a system :

1) Parts of a system sustain the themal stress at operation and such stress accumulates gradually. The damage caused during operation has an identical probability distribution $G(x)$ with finite mean, and each damage is additive. Then, the total damage $Z_N \equiv \sum_{i=1}^{N} Y_i$ until time NT, where $Z_0 \equiv 0$, has a distribution $\Pr\{Z_N \leq x\} = G^{(N)}(x)$ $(N = 1, 2, \cdots)$, where $\Phi^{(N)}(x)$ $(N = 1, 2, \cdots)$ denotes the N-fold Stieltjes convolution of $\Phi(x)$ with itself and $\Phi^{(0)}(x) \equiv 1$ for $x \geq 0$.

2) The PM is performed at times iT $(i = 1, 2, \cdots, N)$ periodically. The system is operating continuously until time $(N+1)T$. The probability that shocks occur during $(iT, (i+1)T]$ $(i = 0, 1, 2, \cdots, N)$ is statistically identical.

3) When the cumulative damage is below a managerial level k during $(iT, (i+1)T]$, the system undergoes the PM at time $(i+1)T$ and its cost is c_p.

4) When the total damage exceeds a failure level K_i $(i = 0, 1, \cdots, N)$, the system fails and its maintenance cost is c_f. The operation time yields profit c_e per hour. It is assumed that K_i declines with operation time because some cumulative damage of parts remains after PM. Although a real K_i declines gradually, an idealized K_i is constant during $(iT, (i+1)T]$ $(i = 0, 1, 2, \cdots, N)$ and it declines stepwise for the simplicity of analysis.
5) When the total damage is between k and K_i, the system is overhauled and its cost is c_o $(c_f > c_o > c_p)$.
6) The system operates until its total damage exceeds a level k, K_i or at time $(N+1)T$, whichever occurs first.

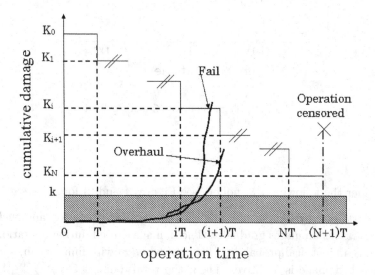

Figure 2. Schematic diagram of operation of a damage system.

The probability p_i that the system undergoes overhaul when the total damage exceeds k at time $(i+1)T$ is

$$p_i = G^{(i)}(k) - G^{(i+1)}(k), \tag{1}$$

the probability p_{K_i} that the system fails during $(iT, (i+1)T]$, when the total damage exceeds K_i, is

$$p_{K_i} = \int_0^k [1 - G(K_i - u)] dG^{(i)}(u), \tag{2}$$

the probability p_T that the system operates at time $(N+1)T$ when the total damage is below k, is

$$p_T = G^{(N+1)}(k). \tag{3}$$

It is obvious that $\sum_{i=0}^{N} p_i + p_T = 1$. Suppose that $G(x) = 1 - e^{-\mu x}$, i.e., $G^{(n)}(x) = \sum_{i=n}^{\infty} (\mu x)^i e^{-\mu x}/i!$ $(n = 0, 1, 2, \cdots)$. The expected profit P_1 until time $(N+1)T$ is, from (1), (2) and (3),

$$\begin{aligned} P_1(N) &= \sum_{i=0}^{N} [c_e(iT+t_0) - c_p i - c_f] p_{K_i} \\ &+ \sum_{i=0}^{N} [c_e(iT+t_0) - c_p i - c_o](p_i - p_{K_i}) \\ &+ [c_e(N+1)T - c_p N - c_o] p_T \\ &= -(c_f - c_o) \sum_{i=0}^{N} [e^{-\mu(K_i - k)} - e^{-\mu(K_{i-1} - k)}] G^{(i)}(k) \\ &+ (c_e T - c_p) \sum_{i=0}^{N} G^{(i)}(k) \\ &+ [(c_f - c_o) e^{-\mu(K_N - k)} + c_e(T - t_0)] G^{(N+1)}(k) \\ &- c_e(T - t_0) - c_o + c_p \quad (N = 0, 1, 2, \cdots), \end{aligned} \tag{4}$$

where $K_i \equiv \infty$ when $i < 0$ and t_0 denotes the mean time the total damage level exceeds k during $(iT, (i+1)T]$.

Forming the inequality $P_1(N+1) - P_1(N) \leq 0$,

$$Q_N(k) \left[e^{-\mu(K_N - k)} + \frac{c_e(T - t_0)}{c_f - c_o} \right] \geq \frac{c_e T - c_p}{c_f - c_o} \quad (N = 0, 1, 2, \cdots), \tag{5}$$

where

$$Q_N(k) \equiv \frac{G^{(N)}(k) - G^{(N+1)}(k)}{G^{(N)}(k)} \quad (i = 0, 1, 2, \cdots).$$

Denoting the left-hand side of (5) by $L_1(N)$,

$$L_1(\infty) \equiv \lim_{N \to \infty} L_1(N) = e^{-\mu(K_\infty - k)} + \frac{c_e(T - t_0)}{c_f - c_o}. \tag{6}$$

Forming the inequality $L_1(N) - L_1(N-1) \geq 0$,

$$\frac{Q_N(k) - Q_{N-1}(k)}{Q_{N-1}(k)} \geq -\frac{e^{-\mu K_N} - e^{-\mu K_{N-1}}}{e^{-\mu K_N} + [c_e(T - t_0)]/(c_f - c_o) e^{-\mu k}}. \tag{7}$$

The right-hand side of (7) is less than zero because $K_{N-1} > K_N$, and

$$Q_N(k) = \frac{1}{\sum_{i=N}^{\infty} \frac{N!(\mu k)^i}{i!(\mu k)^N}} = \frac{1}{1 + \frac{\mu k}{N+1} + \frac{\mu k}{N+1}\frac{\mu k}{N+2} + \cdots}. \tag{8}$$

Thus, $Q_N(k)$ is an increasing function of N, and the left-hand side of (7) is greater than zero. Therefore, $L_1(N)$ is an increasing function of N. Thus, if $L_1(\infty) > (c_e T - c_p)/(c_f - c_o)$, then there exists a unique and finite N^* $(1 \le N^* < \infty)$ which maximizes $P_1(N)$ in (4).

3. Model 2

We consider the following PM policy which has the same assumptions as Model 1 except 2 which is rewritten as :

2') Shocks during operation occur at a non-homogeneous Poisson process. The PM is performed at time iT $(i = 1, 2, \cdots, N)$ periodically and the probability that the n-th shock occurs during $(0, t]$ is $H_n(t) = (\lambda t)^n e^{-\lambda t}/n!$ [18]. Thus, the probability that shocks occur at more than n-times during $(0, t]$ is $F_n(t) \equiv \sum_{i=n}^{\infty} H_i(t)$ $(n = 0, 1, 2, \cdots)$. The system is operating continuously until time $(N+1)T$.

The probabilities p_i, p_{K_i} and p_T are, respectively,

$$p_i = \sum_{n=1}^{\infty} [G^{(n-1)}(k) - G^{(n)}(k)] \int_{iT}^{(i+1)T} dF_n(t), \tag{9}$$

$$p_{K_i} = \sum_{n=1}^{\infty} \int_0^k [1 - G(K_i - u)] dG^{(n-1)}(u) \int_{iT}^{(i+1)T} dF_n(t), \tag{10}$$

$$p_T = \sum_{n=0}^{\infty} G^{(n)}(k) H_n((N+1)T). \tag{11}$$

It is obvious that $\sum_{i=0}^{N} p_i + p_T = 1$.

The expected profit $P_2(N)$ until time $(N+1)T$ is, from (4), (9), (10) and (11),

$$P_2(N) = -(c_f - c_o)\left\{\sum_{i=0}^{N}[e^{-\mu(K_i-k)} - e^{-\mu(K_{i-1}-k)}]\sum_{n=0}^{\infty}G^{(n)}(k)H_n(iT)\right.$$

$$\left. + e^{-\mu(K_N-k)}\sum_{n=0}^{\infty}G^{(n)}(k)H_n((N+1)T)\right\}$$

$$+ c_e T\sum_{i=1}^{N+1}\sum_{n=0}^{\infty}G^{(n)}(k)H_n(iT) - c_p\sum_{i=0}^{N}\sum_{n=0}^{\infty}G^{(n)}(k)H_n(iT)$$

$$- c_e t_0 \sum_{n=0}^{\infty}G^{(n)}(k)H_n((N+1)T)$$

$$+ c_e t_0 - c_o + c_p \quad (N = 0, 1, 2, \cdots). \qquad (12)$$

Forming the inequality $P_2(N+1) - P_2(N) \le 0$,

$$\frac{\sum_{n=0}^{\infty}G^{(n)}(k)[H_n(NT) - H_n((N+1)T)]}{\sum_{n=0}^{\infty}G^{(n)}(k)H_n(NT)}\left[e^{-\mu(K_N-k)} + \frac{c_e(T-t_0)}{c_f - c_o}\right]$$

$$\ge \frac{c_e T - c_p}{c_f - c_o} \quad (N = 0, 1, 2, \cdots). \qquad (13)$$

Denoting the left-hand side of (13) by $L_2(N)$,

$$L_2(\infty) = e^{-\mu(K_\infty - k)} + \frac{c_e(T-t_0)}{c_f - c_o}. \qquad (14)$$

Thus, if $L_2(\infty) > (c_e T - c_p)/(c_f - c_o)$, then there exists a finite N^* which maximizes $P_2(N)$ in (12).

4. Numerical Illustrations

Suppose that $K_i \equiv (K_0 - K_\infty)e^{-\sigma iT} + K_\infty$ because K_i decreases monotonically. Table 1 gives the optimal PM number N_1^* and the maximum expected profit $P_1(N_1^*)$ until the censoring of operation of Model 1, when $\mu = 0.1, 0.12, 0.14$, $c_e = 10, 5, 1$, $c_p = 1, 100, 200$, $c_o = 50, 150, 250$, $c_f = 1000, 1500, 3000$, $k = 90, 100, 110$, $T = 500$, $t_0 = 250$, $K_0 = 100$, $K_\infty = 50$ and $\sigma = 5 \times 10^{-5}$, and $t_0 = T/2$ because the occurrence probability of shocks is assumed to be statistically identical. In this illustration, N_1^* increases when μ, $1/c_e$, $1/c_o$, c_f and k decrease. It is of interest that N_1^* does not change when c_p changes.

Table 2 gives the optimal PM time N_2^* and the maximum expected profit $P_2(N_2^*)$ of Model 2, when $\lambda = 1 \times 10^{-3}, 9 \times 10^{-4}, 8 \times 10^{-4}$, $\mu =$

0.1, 0.12, 0.14, $c_e = 10, 5, 1$, $c_p = 1, 100, 200$, $c_o = 50, 150, 250$, $c_f = 1000, 1500, 3000$, $k = 90, 100, 110$, $T = 500$, $t_0 = 250$, $K_0 = 100$, $K_\infty = 50$ and $\sigma = 5 \times 10^{-5}$. In this illustration, N_2^* increases when λ and c_p decreases and other variations of N_2^* are the same as ones in Table 1.

In these illustrations, N_2^* and $P_2(N_2^*)$ are greater than N_1^* and $P_1(N_1^*)$ respectively. When the change rate of N_1 with c_e is denoted as $N_1'(c_e)(\equiv |\Delta N_1/\Delta \mu|)$, $N_2'(c_e)/N_1'(c_e)$, $N_2'(c_o)/N_1'(c_o)$, $N_2'(c_f)/N_1'(c_f)$ and $N_2'(k)/N_1'(k)$ change from 2.4 to 2.7. It is of interest that $N_2'(\mu)/N_1'(\mu)$ is 6. Variations of N_1^* and N_2^* caused by λ, μ, c_e, c_p, c_o, c_f and k in Tables 1 and 2 consistent with the practical experience.

The operation and failure data of power plants which are settled around the world are accumulated systematically by the remote monitoring system [19]. The damage assessment and life estimation of failed units are performed accurately by utilizing a digitized microscope, a computer image processor, and software [3]. Depending on the operation data and damage assessment data, the optimized maintenance plan which minimizes the expenditure throughout the total life of system, is established by using the probabilistic life estimation method [20]. These two models can contribute to establish much efficient maintenance plan.

Table 1. Optimal PM time N_1^* and maximum expected profit $P_1(N_1^*)$ of model 1.

μ	c_e	c_p	c_o	c_f	k	N_1^*	$P_1(N_1^*) \times 10^2$
0.1	10	1	50	1000	90	28	415
0.12	10	1	50	1000	90	27	502
0.14	10	1	50	1000	90	26	588
0.1	5	1	50	1000	90	21	202
0.1	1	1	50	1000	90	11	33
0.1	10	100	50	1000	90	28	407
0.1	10	200	50	1000	90	28	399
0.1	10	1	150	1000	90	30	415
0.1	10	1	250	1000	90	31	415
0.1	10	1	50	1500	90	24	409
0.1	10	1	50	3000	90	21	404
0.1	10	1	50	1000	100	19	445
0.1	10	1	50	1000	110	14	441

5. Conclusions

We have considered the optimal operation censoring policies for a system with cumulative damage. The system fails when the total damage exceeds

Table 2. Optimal PM time N_2^* and maximum expected profit $P_2(N_2^*)$ of model 2.

$\lambda \times 10^{-3}$	μ	c_e	c_p	c_o	c_f	k	N_2^*	$P_1(N_2^*) \times 10^2$
1	0.1	10	1	50	1000	90	66	972
0.9	0.1	10	1	50	1000	90	71	1078
0.8	0.1	10	1	50	1000	90	78	1209
1	0.12	10	1	50	1000	90	54	1131
1	0.14	10	1	50	1000	90	47	1271
1	0.1	5	1	50	1000	90	48	472
1	0.1	1	1	50	1000	90	25	77
1	0.1	10	100	50	1000	90	65	952
1	0.1	10	200	50	1000	90	64	933
1	0.1	10	1	150	1000	90	70	974
1	0.1	10	1	250	1000	90	74	976
1	0.1	10	1	50	1500	90	54	957
1	0.1	10	1	50	3000	90	47	943
1	0.1	10	1	50	1000	100	44	1014
1	0.1	10	1	50	1000	110	31	979

a certain critical level and the critical level lowers at every PM. The overhaul is performed when the total damage is between a managerial level and a critical one. Two models are considered and expected profits of these models are derived. Optimal policies which maximize these two profits are discussed and numerical examples have been given. These policies are useful for considering the cost-optimal operation plan of aged fossil-fired power plants. Abundant operation data of these system can determine various parameters such as failure levels, the damage distribution and the shock occurrence distribution precisely. We can decide the optimal operation censoring time from the economical view point by estimating these parameters.

References

1. K. Hisano, Preventive maintenance and residual life evaluation technique for power plant (I.Preventive Maintenance) (in Japanese), *The Thermal and Nuclear Power* **51** (4), pp.491–517, (2000).
2. K. Hisano, Preventive maintenance and residual life evaluation technique for power plant (I.Preventive Maintenance) (in Japanese), *The Thermal and Nuclear Power* **51** (8), pp.81–101, (2000).
3. T. Kuroishi, Y. Minami, Y. Kobayashi, T. Yokoyama, Y. Hasegawa and M. Minatomoto, Power systems : A portal to customer services for electric power generation, *Mitsubishi Heavy Industries,LTD. Technical Review* **40** (2), pp.1–9, (2003).

4. R. E. Barlow and F. Proschan, *Mathematical Theory of Reliability*. (John Wiley & Sons, New York, 1965).
5. D. R. Cox, *Renewal Theory*. (Methuen, London, 1962).
6. T. Nakagawa, *Shock and Damage Models in Reliability Theory*, (Springer Verlag, London, 2007).
7. J. D. Esary, A. W. Marshall and F. Proschan, Shock models and wear processes, *Annals of Probability*, **1**, pp.627-649, (1973).
8. H. M. Taylor, Optimal replacement under additive damage and other failure models, *Naval Res. Logist. Quart*, **22**, pp.1–18, (1975).
9. T. Nakagawa, A summary of discrete replacement policies, *European J. of Operational Research*, **17**, pp.382–392, (1984).
10. C. Qian, S. Nakamura and T. Nakagawa, Replacement and minimal repair policies for a cumulative damage model with maintenance, *Computers and Mathematics with Applications*, **46**, pp.1111–1118, (2003).
11. R. M. Feldman, Optimal replacement with semi-Markov shock models, *Journal of Applied Probability*, **13**, pp.108–117, (1976).
12. T. Nakagawa, On a replacement problem of a cumulative damage model, *Operational Research Quarterly*, **27**, pp.895–900, (1976).
13. T. Nakagawa and M. Kijima, Replacement policies for a cumulative damage model with minimal repair at failure, *IEEE Trans. Reliability*, **38**, pp.581–584, (1989).
14. W. Nelson, *Accelerated Testing*, (Wiley, New York, 1990).
15. W. Nelson, Prediction of field reliability of units, each under differing dynamic stresses, from accelerated test data, *Handbook of Statistics*, **20** (N. Balakrishnan and C. R. Rao, eds.) pp.611–621, (2001).
16. L. A. Escobar and W. Meeker, A review of accelerated test models, *Statistical Science*, **21** (4), pp.552–577, (2006).
17. S. Kosugiyama, T. Takizuka, K. Kunitomi, X. Yan, S. Katanishi and S. Takada, Basic policy of maintenance for the power conversion system of the gas turbine high temperature reactor 300 (GTHTR300) (in Japanese), *Journal of Nuclear Science and Technology*, **2** (3), pp.105–117, (2003).
18. S. Osaki, *Applied Stochastic Systems Modeling*. (Springer Verlag, Berlin, 1992).
19. T. Kuroishi, N. Osaki, M. Kobayashi, E. Sano, M. Inoue and M. Torichigai, New service business for power industries, *Mitsubishi Heavy Industries, Ltd. Technical Review* **44** (4), pp.45–48, (2007).
20. H. Matsumoto, N.Nishimura, Y. Chuman, S. Kumano, M. Kobayashi, K. Tominaga, K. Hayashi and A. Shibashi, Development of maintenance optimization system for aged fossil fired power plants, *Mitsubishi Heavy Industries, Ltd. Technical Review* **41** (1), pp.14–15, (2004).

Part D Dependable Computing

OPTIMAL SEQUENTIAL CHECKPOINT INTERVALS FOR ERROR DETECTION

KENICHIRO NARUSE

Information Center, Nagoya Sangyo University
3255-5 Arai-cho, Owariasahi 488-8711, Japan
narunaru@gmail.com

TOSHIO NAKAGAWA

Department of Marketing and Information Systems,
Aichi Institute of Technology
1247 Yachigusa, Yakusa-cho, Toyota 470-0392, Japan
toshi-nakagawa@aitech.ac.jp

SAYORI MAEJI

Institute of Consumer Sciences and Human Life
Kinjo Gakuin University
1723 Oomori 2, Moriyama, Nagoya 463-8521, Japan

This paper adopts a modular redundant system as the recovery techniques of error detection and error masking on the finite process execution: Checkpoints are placed at sequential times $T_k(k=1, 2, \cdots, N)$. We consider two checkpoint models where error rates during the interval $(T_{k-1}, T_k)(k=1, 2, \cdots, N)$ increase with the number of checkpoints and with the original execution time. The mean times to completion of the process are obtained analytically, and optimal checkpoint intervals which minimize them are computed numerically by solving simultaneous equations. Furthermore, approximate checkpoint intervals are derived by denoting that the probability of the occurrence of errors during $(T_{k-1}, T_k]$ is constant. It is shown that the approximate method is simple and these intervals give good approximations to optimal ones.

1. Introduction

In the process of computer systems, some errors often occur due to noises, human errors, hardware faults and so on. To detect and mask such errors, some fault tolerant computing techniques have been considered [1, 2]. The simplest scheme in recovery techniques of error detection of the process is as follows: We execute two independent modules which compare two results at checkpoint times. If their results do not match with each other, we go back to the newest checkpoint and make a retrial of the processes. The performance and reliability of a duplex system with a spare processor was evaluated [3]. Furthermore, the performance of checkpoint schemes with task duplication was evaluated [4, 5]. The optimal instruction-retry period which minimizes the probability of the dynamic failure by a triple modular controller was derived [6]. We have already considered a majority decision system as an error masking system, and compared the mean time to completion of the process and decided numerically what a majority decision system is optimal [7, 8].

This paper considers a general modular system of error detection and error masking on a finite process execution with time S: Suppose that checkpoints are placed at sequential times $T_k(k=1, 2, \cdots, N)$, where $T_N \equiv S$. First, it is assumed that error rates during the interval $(T_{k-1}, T_k]$ ($k=1,2, \cdots, N$) increase with the number k of checkpoints. The mean time to completion of the process is obtained, and optimal checkpoint intervals which minimize it are derived by solving simultaneous equations. Furthermore, approximate checkpoint intervals are given by denoting that the probability of the occurrence of errors during $(T_{k-1}, T_k]$ is constant. Secondly, it is assumed that error rates during $(T_{k-1}, T_k]$ increase with the original execution time, irrespective of the number of recoveries. Optimal checkpoint intervals which minimize the mean time to completion of the process are discussed, and their approximate times are shown. Numerical examples of optimal checkpoint times for a double modular system are presented. It is shown numerically that the approximate method is simple and these intervals give good approximations to optimal ones.

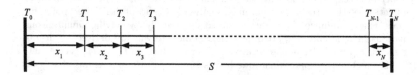

Figure 1. Sequential checkpoint interval.

2. Sequential Checkpoint Interval

Suppose that S is a native execution time of the process which does not include the overheads of retries and checkpoint generations. Then, we divide S into N parts and create a checkpoint at sequential times $T_k(k = 1, 2, \cdots, N-1)$, where $T_0 \equiv 0$ and $T_N \equiv S$ (Figure 1).

Let us introduce a constant overhead C for the comparison of a modular system. Further, the probability that a modular system has no error during the interval $(T_{k-1}, T_k]$ is $\overline{F}_k(T_k - T_{k-1})$ irrespective of other intervals and rollback operation. Then, the mean time $L_1(N)$ to completion of the process is the summation of the processing times and the overhead C for the comparison of a modular system. From the assumption that a modular system is rolled back to the previous checkpoint when some error has been detected at a checkpoint, the mean execution time of the process for the interval $(T_{k-1}, T_k]$ is given by renewal equation

$$L_1(k) = (T_k - T_{k-1} + C)\overline{F}_k(T_k - T_{k-1})$$
$$+ (T_k - T_{k-1} + C + L_1(k))F_k(T_k - T_{k-1}), \qquad (1)$$

and solving it,

$$L_1(k) = \frac{T_k - T_{k-1} + C}{\overline{F}_k(T_k - T_{k-1})} \qquad (k = 1, 2, \cdots N), \qquad (2)$$

where $F_k(t) \equiv 1 - \overline{F}_k(t)$. Thus, the mean time to completion of the process is

$$L_1(N) \equiv \sum_{k=1}^{N} L_1(k) = \sum_{k=1}^{N} \frac{T_k - T_{k-1} + C}{\overline{F}_k(T_k - T_{k-1})} \qquad (N = 1, 2, \cdots). \qquad (3)$$

We find optimal times T_k which minimize $L_1(N)$ for a specified N. Let $f_k(t)$ be a density function of $F_k(t)$ and $r_k(t) \equiv f_k(t)/\overline{F}_k(t)$ that is the failure rate of $F_k(t)$. Then, differentiating $L_1(N)$ with respect to T_k and setting it equal to zero,

$$\frac{1}{\overline{F}_k(T_k - T_{k-1})}\left[1 + (T_k - T_{k-1} + C)\, r_k(T_k - T_{k-1})\right]$$
$$= \frac{1}{\overline{F}_{k+1}(T_{k+1} - T_k)}\left[1 + (T_{k+1} - T_k + C)\, r_{k+1}(T_{k+1} - T_k)\right]. \qquad (4)$$

Setting that $x_k \equiv T_k - T_{k-1}$ and rewriting (4) as a function of x_k,

$$\frac{1}{\overline{F}_k(x_k)}\left[1 + (x_k + C)\, r_k(x_k)\right]$$
$$= \frac{1}{\overline{F}_{k+1}(x_{k+1})}\left[1 + (x_{k+1} + C)\, r_{k+1}(x_{k+1})\right] \qquad (k = 1, 2, \cdots N - 1). \qquad (5)$$

Next, suppose that $\overline{F}_k(t) = e^{-\lambda_k t}$, i.e., an error rate during $(T_{k-1}, T_k]$ is constant λ_k which increases with k. Then, Equation (5) is rewritten as

$$\frac{1+\lambda_{k+1}(x_{k+1}+C)}{1+\lambda_k(x_k+C)} - e^{(\lambda_k x_k - \lambda_{k+1} x_{k+1})} = 0. \tag{6}$$

If $\lambda_{k+1} x_{k+1} > \lambda_k x_k$, then the first time in (6) is greater than 1, however, the second time is less than 1. Thus, the condition that there exists some solution of (6) is $\lambda_{k+1} x_{k+1} \leq \lambda_k x_k$, and hence, $x_{k+1} \leq x_k$ since $\lambda_{k+1} \geq \lambda_k$.

In particular, when $\lambda_k \equiv \lambda$ for $k = 1, 2, \cdots N$, Equation (6) becomes

$$\frac{1+\lambda(x_{k+1}+C)}{1+\lambda(x_k+C)} - e^{\lambda(x_k - x_{k+1})} = 0. \tag{7}$$

Since $x_{k+1} \leq x_k$, we have that $x_{k+1} \geq x_k$ from (7), i.e., it is easily proved that a solution to satisfy (7) is restricted only to $x_{k+1} = x_k \equiv T$, irrespective of the interval number k. Then, the mean time to completion of the process is

$$L_1(N) = (S + NC) e^{\lambda S/N}. \tag{8}$$

The optimal number N^* and time $T^* = S/N^*$ which minimize $L_1(N)$ were discussed analytically and numerically [6].

If $\lambda_{k+1} > \lambda_k$, then $x_{k+1} < x_k$ from (6). Let $Q(x_{k+1})$ be the left-hand side of (6) for a fixed x_k. Then, $Q(x_{k+1})$ is strictly increasing from

$$Q(0) = \frac{1+\lambda_{k+1}C}{1+\lambda_k(x_k+C)} - e^{\lambda_k x_k}$$

to $Q(x_k) > 0$. Thus, if $Q(0) < 0$, then an optimal $x^*_{k+1} (0 < x^*_{k+1} < x_k)$ to satisfy (6) exists uniquely.

Therefore, noting that $T_0 = 0$ and $T_N = S$, we have the following result:

(i) When $N = 1$ and $T_1 = S$, the mean time is

$$L_1(1) = (S + C) e^{\lambda_1 S}. \tag{9}$$

(ii) When $N = 2$, from (6),

$$[1 + \lambda_1 (x_1 + C)] e^{\lambda_1 x_1} - [1 + \lambda_2 (S - x_1 + C)] e^{\lambda_2 (S - x_1)} = 0. \tag{10}$$

Letting $Q_1(x_1)$ be the left-hand side of (10), it is strictly increasing from $Q_1(0) < 0$ to

$$Q_1(S) = [1 + \lambda_1 (S + C)] e^{\lambda_1 S} - (1 + \lambda_2 C).$$

Hence, if $Q_1(S) > 0$, then $x_1^* = T_1^* (0 < T_1^* < S)$ to satisfy (10) exists uniquely, and conversely, if $Q_1(S) \leq 0$ then $x_1^* = T_1^* = S$.

(iii) When $N = 3$, we compute $x_k^*(k = 1, 2)$ which satisfy the simultaneous equations:

$$[1 + \lambda_1 (x_1 + C)] e^{\lambda_1 x_1} = [1 + \lambda_2 (x_2 + C)] e^{\lambda_2 x_2}, \tag{11}$$

$$[1 + \lambda_2 (x_2 + C)] e^{\lambda_2 x_2} = [1 + \lambda_3 (S - x_1 - x_2)] e^{\lambda_3 (S - x_1 - x_2)}. \tag{12}$$

(iv) When $N = 4, 5, \cdots$, we compute x_k^* and $T_k = \sum_{j=1}^{k} x_j^*$ similarly.

We compute sequential checkpoint intervals $T_k (k = 1, 2, \cdots, N)$ for a double modular system. It is assumed that $\lambda_k = 2[1 + \alpha(k - 1)]\lambda$ ($k = 1, 2, \cdots$), i.e., an error rate increases by 100% of an original rate λ of one module. Table 1 presents optimal sequential intervals λT_k and the resulting mean times $\lambda L_1(N)$ for $N = 1, 2, \cdots, 9$ when $\alpha = 0.1$, $\lambda S = 10^{-1}$ and $\lambda C = 10^{-3}$. In this case, the mean time is the smallest when $N = 5$, i.e., the optimal checkpoint number is $N^* = 5$ and the checkpoint times $T_k^*(k = 1, 2, 3, 4, 5)$ should be placed at 2.38, 4.53, 6.50, 8.32, 10.00(sec) for $\lambda = 10^{-2}(1/\text{sec})$, and the mean time 11.009 is about 10% longer than an original execution time $S = 10$. Further, all values of $x_k = T_k - T_{k-1}$ decrease with k because error rates increase with the number of checkpoints.

It is very troublesome to solve simultaneous equations. We consider the following approximate checkpoint times: It is assumed that the probability that a modular system has no error during $(T_{k-1}, T_k]$ is constant, i.e., $\overline{F}_k(T_k - T_{k-1}) \equiv q(k = 1, 2, \cdots, N)$. From this assumption, we derive $T_k - T_{k-1} \equiv \overline{F}_k^{-1}(q)$ as a function of q. Substituting this $T_k - T_{k-1}$ into (3), the mean time to completion of the process is

$$L_1(N) = \sum_{k=1}^{N} \frac{\overline{F}_k^{-1}(q) + C}{q}. \tag{13}$$

We discuss an optimal q which minimizes $L_1(N)$. For example, when $\overline{F}_k(t) = e^{-\lambda_k t}$,

$$e^{-\lambda_k (T_k - T_{k-1})} = q \equiv e^{-\tilde{q}},$$

and hence,

$$T_k - T_{k-1} = \frac{\tilde{q}}{\lambda_k}.$$

Since

$$\sum_{k=1}^{N} (T_k - T_{k-1}) = T_N = S = \tilde{q} \sum_{k=1}^{N} \frac{1}{\lambda_k},$$

we have

$$L_1(N) = e^{\tilde{q}}\left[\tilde{q}\sum_{k=1}^{N}\frac{1}{\lambda_k} + NC\right] = e^{\tilde{q}}(S + NC). \quad (14)$$

Therefore, we compute $\tilde{q}$ and $L_1(N)$ for a specified N. Comparing $L_1(N)$ for $N = 1, 2, \cdots$, we obtain an optimal $\widetilde{N}$ which minimizes $L_1(N)$ and $\tilde{q} = S/\sum_{k=1}^{N}(1/\lambda_k)$. Lastly, we may compute $\widetilde{T}_k = \tilde{q}\sum_{j=1}^{k}(1/\lambda_j)(k = 1, 2, \cdots, \widetilde{N} - 1)$ for an approximate optimal $\widetilde{N}$ which minimizes $L_1(N)$.

Table 2 presents $\tilde{q} = S/\sum_{k=1}^{N}(1/\lambda_k)$ and $\lambda L_1(N)$ in (14) for $N = 1, 2, \cdots, 9$ under the same assumptions as those in Table 1. In this case, $\widetilde{N} = 5 = N^*$ and the mean time $L_1(5)$ is a little longer than that in Table 1. When $\widetilde{N} = 5$, approximate checkpoint times are $\lambda\widetilde{T}_k \times 10^2$=2.37, 4.52, 6.49, 8.31, 10.00 that are a little shorter than those in Table 1. Such computations

Table 1. Checkpoint intervals λT_k and mean time $\lambda L_1(N)$ when $\lambda_k = 2[1+ 0.1(k-1)]\lambda$, $\lambda S = 10^{-1}$ and $\lambda C = 10^{-3}$.

N	1	2	3	4	5
$\lambda T_1 \times 10^2$	10.00	5.24	3.65	2.85	2.38
$\lambda T_2 \times 10^2$		10.00	6.97	5.44	4.53
$\lambda T_3 \times 10^2$			10.00	7.81	6.50
$\lambda T_4 \times 10^2$				10.00	8.32
$\lambda T_5 \times 10^2$					10.00
$\lambda L_1(N) \times 10^2$	12.33617	11.32655	11.07923	11.00950	11.00887

N	6	7	8	9
$\lambda T_1 \times 10^2$	2.05	1.83	1.65	1.52
$\lambda T_2 \times 10^2$	3.91	3.48	3.15	2.89
$\lambda T_3 \times 10^2$	5.62	4.99	4.52	4.15
$\lambda T_4 \times 10^2$	7.19	6.39	5.78	5.31
$\lambda T_5 \times 10^2$	8.65	7.68	6.95	6.38
$\lambda T_6 \times 10^2$	10.00	8.88	8.03	7.37
$\lambda T_7 \times 10^2$		10.00	9.05	8.31
$\lambda T_8 \times 10^2$			10.00	9.18
$\lambda T_9 \times 10^2$				10.00
$\lambda L_1(N) \times 10^2$	11.04228	11.09495	11.15960	11.23220

are much easier than to solve simultaneous equations. It would be sufficient to adopt approximate checkpoint intervals as optimal ones in actual fields. Figure 2 draws the mean time $\lambda L_1(N)$ for $1 \leq N \leq 20$.

Table 2. Mean time $\lambda L_1(N)$ for $\widetilde{q}$ when $\lambda S = 10^{-1}$ and $\lambda C = 10^{-3}$.

N	$\widetilde{q}$	$\lambda L_1(N) \times 10^2$
1	0.2000000	12.33617
2	0.1047619	11.32655
3	0.0729282	11.07923
4	0.0569532	11.00951
5	0.0473267	11.00888
6	0.0408780	11.04229
7	0.0362476	11.09496
8	0.0327555	11.15962
9	0.0300237	11.23222

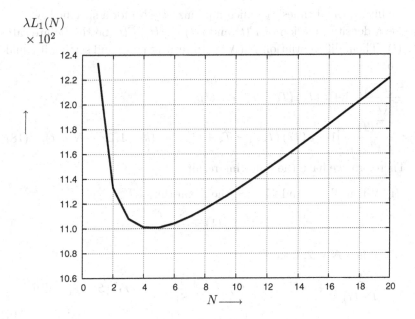

Figure 2. Mean time $\lambda L_1(N)$ when $1 \leq N \leq 20$.

3. Model 2

It has been assumed until now that error rates increase with the number of checkpoints. We assume for the simplicity of the model that the probability that a modular system has no error during the interval $(T_{k-1}, T_k]$ is $\overline{F}(T_k)/\overline{F}(T_{k-1})$, irrespective of rollback operation. Then, the mean execution time of the process for the interval $(T_{k-1}, T_k]$ is given by a renewal equation

$$L_2(k) = (T_k - T_{k-1} + C)\frac{\overline{F}(T_k)}{\overline{F}(T_{k-1})}$$
$$+ (T_k - T_{k-1} + C + L_2(k))\frac{F(T_k) - F(T_{k-1})}{\overline{F}(T_{k-1})}, \quad (15)$$

and solving it,

$$L_2(k) = \frac{(T_k - T_{k-1} + C)\overline{F}(T_{k-1})}{\overline{F}(T_k)} \quad (k = 1, 2, \cdots, N). \quad (16)$$

Thus, the mean time to completion of the process is

$$L_2(N) = \sum_{k=1}^{N} \frac{(T_k - T_{k-1} + C)\overline{F}(T_{k-1})}{\overline{F}(T_k)} \quad (N = 1, 2, \cdots). \quad (17)$$

We find optimal times T_k which minimize $L_2(N)$ for a specified N. Let $f(t)$ be a density function of $F(t)$ and $r(t) \equiv f(t)/\overline{F}(t)$ be the failure rate of $F(t)$. Then, differentiation $L_2(N)$ with respect to T_k and setting it equal to zero,

$$\frac{\overline{F}(T_{k-1})}{\overline{F}(T_k)}[1 + r(T_k)(T_k - T_{k-1} + C)]$$
$$= \frac{\overline{F}(T_k)}{\overline{F}(T_{k+1})}[1 + r(T_k)(T_{k+1} - T_k + C)] \quad (k = 1, 2, \cdots N-1). \quad (18)$$

Therefore, we have the following result:

(i) When $N = 1$ and $T_1 = S$, the mean time is

$$L_2(1) = \frac{S + C}{\overline{F}(S)}. \quad (19)$$

(ii) When $N = 2$, from (18)

$$\frac{1}{\overline{F}(T_1)}[1 + r(T_1)(T_1 + C)] - \frac{\overline{F}(T_1)}{\overline{F}(S)}[1 + r(T_1)(S - T_1 + C)] = 0. \quad (20)$$

Letting $Q_2(T_1)$ be the left-hand side of (20), it is evidently seen that

$$Q_2(0) = 1 + r(0)C - \frac{1}{\overline{F}(S)}[1 + r(0)(S+C)] < 0,$$

$$Q_2(S) = \frac{1}{\overline{F}(S)}[1 + r(S)(S+C)] - [1 + r(S)C] > 0.$$

Thus, there exists a T_1 that satisfies (20).

(iii) When $N = 3$, we compute T_k $(k = 1, 2)$ which satisfy the simultaneous equations:

$$\frac{1}{\overline{F}(T_1)}[1 + r(T_1)(T_1+C)] = \frac{\overline{F}(T_1)}{\overline{F}(T_2)}[1 + r(T_1)(T_2-T_1+C)], \tag{21}$$

$$\frac{\overline{F}(T_1)}{\overline{F}(T_2)}[1 + r(T_2)(T_2-T_1+C)] = \frac{\overline{F}(T_2)}{\overline{F}(S)}[1 + r(T_2)(S-T_2+C)]. \tag{22}$$

(iv) When $N = 4, 5, \cdots$, we compute T_k similarly.

We compute sequential checkpoint intervals T_k ($k = 1, 2, \cdots, N$) when error rates increase with the original execution time. It is assumed that $\overline{F}(t) = e^{-2(\lambda t)^m}$ ($m > 1$), $\lambda C = 10^{-3}$ and $\lambda S = 10^{-1}$. Table 3 presents optimal sequential intervals λT_k and the resulting mean times $\lambda L_2(N)$ for $N = 1, 2, \cdots, 9$ when $\overline{F}(t) = \exp[-2(\lambda t)^{1.1}]$, $\lambda S = 10^{-1}$ and $\lambda C = 10^{-3}$. In this case, the mean time is the smallest when $N = 4$, i.e., the optimal checkpoint number is $N^* = 4$ and the checkpoint times T_k^* ($k = 1, 2, 3, 4$) should be placed at 2.67, 5.17, 7.60, 10.00 (sec) for $\lambda = 10^{-2}(1/\text{sec})$, and the mean time 10.8207 is about 8% longer than an original execution time $S = 10$.

Next, we consider the approximate method similar to that of the previous model. It is assumed that the probability that a modular system has no error during $(T_{k-1}, T_k]$ is constant, i.e., $\overline{F}(T_k)/\overline{F}(T_{k-1}) = q$ ($k = 1, 2, \cdots, N$). When $\overline{F}(t) = e^{-2(\lambda t)^m}$,

$$\frac{\overline{F}(T_k)}{\overline{F}(T_{k-1})} = e^{-2[(\lambda T_k)^m - (\lambda T_{k-1})^m]} = q \equiv e^{-\tilde{q}},$$

and hence,

$$2(\lambda T_k)^m - 2(\lambda T_{k-1})^m = \tilde{q} \qquad (k = 1, 2, \cdots, N).$$

Table 3. Checkpoint intervals when $\overline{F}(t) = \exp[-2\,(\lambda t)^{1.1}]$, $\lambda C = 10^{-3}$ and $\lambda S = 10^{-1}$.

N	1	2	3	4	5
$\lambda T_1 \times 10^2$	10.00	5.17	3.51	2.67	2.16
$\lambda T_2 \times 10^2$		10.00	6.80	5.17	4.18
$\lambda T_3 \times 10^2$			10.00	7.60	6.15
$\lambda T_4 \times 10^2$				10.00	8.09
$\lambda T_5 \times 10^2$					10.00
$\lambda L_2(N) \times 10^2$	11.83902	11.04236	10.85934	10.82069	10.83840

N	6	7	8	9
$\lambda T_1 \times 10^2$	1.81	1.57	1.38	1.23
$\lambda T_2 \times 10^2$	3.51	3.03	2.67	2.39
$\lambda T_3 \times 10^2$	5.17	4.46	3.93	3.51
$\lambda T_4 \times 10^2$	6.80	5.87	5.17	4.62
$\lambda T_5 \times 10^2$	8.41	7.26	6.39	5.71
$\lambda T_6 \times 10^2$	10.00	8.63	7.60	6.80
$\lambda T_7 \times 10^2$		10.00	8.81	7.87
$\lambda T_8 \times 10^2$			10.00	8.94
$\lambda T_9 \times 10^2$				10.00
$\lambda L_2(N) \times 10^2$	10.88391	10.94517	11.01622	11.09376

Table 4. Mean time $\lambda L_2(N)$ for $\widetilde{q}$ when $\lambda S = 10^{-1}$ and $\lambda C = 10^{-3}$.

N	$\widetilde{q}$	$\lambda L_2(N) \times 10^2$
1	0.1588656	11.83902
2	0.0794328	11.04326
3	0.0529552	10.86014
4	0.0397164	10.82136
5	0.0317731	10.83897
6	0.0264776	10.88441
7	0.0226951	10.94561
8	0.0198582	11.01661
9	0.0176517	11.09411

Thus,
$$(\lambda T_k)^m = \frac{k\tilde{q}}{2},$$

i.e.,
$$\lambda T_k = \left(\frac{k\tilde{q}}{2}\right)^{1/m} \quad (k=1,2,\cdots,N-1),$$

and
$$\lambda T_N = \lambda S = \left(\frac{N\tilde{q}}{2}\right)^{1/m}.$$

Therefore,
$$L_2(N) = e^{\tilde{q}}(S+NC) = e^{2(\lambda S)^m/N}(S+NC). \tag{23}$$

Forming the inequality $L_2(N+1) - L_2(N) \geq 0$,
$$C \geq (S+NC)\{e^{2(\lambda S)^m/[N(N+1)]} - 1\}. \tag{24}$$

It is easily proved that the right-hand side of (24) is strictly decreasing to 0. Thus, an optimal $\tilde{N}$ to minimize $L_2(N)$ in (23) is given by a unique minimum which satisfies (24).

Table 4 presents $\tilde{q} = 2(\lambda S)^m/N$ and $\lambda L_2(N)$ in (23) for $N = 1, 2, \cdots, 9$ under the same assumptions in Table 3. In this case, $\tilde{N} = 4 = N^*$ and approximate checkpoint times are $\lambda \tilde{T}_k \times 10^2 = 2.84, 5.33, 7.70, 10.00$, that are a little longer than those of Table 3.

4. Conclusions

We have considered two checkpoint models with a finite execution time S where error rates increase with the number of checkpoints and with the original execution time. The mean times to completion of the process for two models have been obtained and the computing procedures for determining optimal checking intervals to minimize them have been shown. When error rates have an exponential and a Weibull distributions, sequential checkpoint intervals have been computed numerically by solving simultaneous equations. Furthermore, approximate checkpoint intervals have been derived by assuming that the probability of the occurrence of errors during each checkpoint interval is constant. This is very simple and gives good approximations to optimal intervals. It would be sufficient to use practically approximate checkpoint intervals for actual models.

References

1. K. M. Chandy and C. V. Ramamoorthy, Rollback and recovery strategies for computer programs, *IEEE Transactions on Computers*, **21**, 6, 546–556(1972).
2. T. Anderson, and P. Lee, *Principles and Practice*, Prentice Hall, New Jersey(1981).
3. D. K. Pradham and N. H. Vaidya, Roll-forward and rollback recovery: Performance-reliability trade-off, *Proceeding of the 24th International Symposium on Fault-Tolerant Computings*, 186–195(1994).
4. A. Ziv and J. Bruck, Performance optimization of checkpointing schemes with task duplication, *IEEE Transactions on Computers*, **46**, 12, 1381–1386(1997).
5. A. Ziv and J. Bruck, Analysis of checkpointing schemes with task duplication, *IEEE Transactions on Computers*, **47**, 2, 222–227(1998).
6. S. Nakagawa, S. Fukumoto and N. Ishii, Optimal checkpoint interval for redundant error detection and masking systems, *Proceeding of the First Euro-Japanese Workshop on Stochastic Risk Modeling for Finance, Insurance*, Production and Reliability, vol. II(1998).
7. S. Nakagawa, S. Fukumoto and N. Ishii, Optimal checkpointing intervals of three error detection schemes by a double modular redundancy, *Mathematical and Computing Modeling*, **38**, 11–13, 1357–1363(2003).
8. K. Naruse, T. Nakagawa and S. Maeji, Optimal checkpoint intervals for error detection by multiple modular redundancies, *Advanced Reliability Modeling II*, 293–300(2006).

EFFECTIVE ALGORITHMS TO ESTIMATE THE OPTIMAL SOFTWARE REJUVENATION SCHEDULE UNDER CENSORING

KOICHIRO RINSAKA AND TADASHI DOHI

Faculty of Business Administration,
Kobe Gakuin University,
1-3-1 Minatojima, Chuo-ku, Kobe 658-8586, Japan,
rinsaka@ba.kobegakuin.ac.jp

Department of Information Engineering,
Graduate School of Engineering,
Hiroshima University, Higashi-Hiroshima 739-8527, Japan
dohi@rel.hiroshima-u.ac.jp

In this chapter, we consider the optimal software rejuvenation schedule which maximizes the steady-state system availability. We develop statistical algorithms to improve the estimation accuracy in the situation where randomly censored failure time data are obtained. More precisely, based on the kernel density estimation, we estimate the underlying failure time distribution. We propose the framework based on the kernel density estimation to estimate optimal software rejuvenation schedules from censored sample data. In simulation experiments, we show the improvement in the convergence speed to the real optimal solution in comparison with the conventional algorithm.

1. Introduction

Present day applications impose stringent requirements in terms of software dependability since in many cases the consequences of software failure can lead to a huge economic loss or risk to human life. However, these requirements are very difficult to design for and guarantee, particularly in applications of non-trivial complexity. In recent years, considerable attention has been devoted to continuously running software systems whose performance characteristics are smoothly degrading in time. When software application executes continuously for long periods of time, some of the faults cause software to age due to the error conditions that accrue with time and/or load. This phenomenon is called *software aging* and can be observed in many real software systems[1,2,3].

Common experience suggests that most software failures are transient in nature[3]. Since transient failures will disappear if the operation is retried later in slightly different context, it is difficult to characterize their root origin. Therefore, the residual faults have to be tolerated in the operational phase. A novel approach to handle transient software failures is called *software rejuvenation* which can be regarded as a preventive and proactive solution that is particularly useful for counteracting the phenomenon of software aging. It involves stopping the running software occasionally, cleaning its internal state and restarting it. Cleaning the internal state of software might involve garbage collection, flushing operating system kernel tables, reinitializing internal data structures, etc. An extreme, but well known example of rejuvenation which has been around as long as computers themselves is a hardware reboot.

Huang et al.[4] consider a continuous time Markov chain (CTMC) with four states, *i.e.*, initial robust (clean), failure probable, rejuvenation and failure states. They evaluate both the unavailability and the operating cost in steady state under the random software rejuvenation schedule. Dohi et al.[5,6] extend the result of Huang et al.[4] and propose the software rejuvenation model based on a semi-Markov process. They propose non-parametric estimation algorithms based on the empirical distribution to obtain the optimal software rejuvenation schedule from the complete sample of failure time data. If a lot of sample of failure time data can be obtained, then with probability 1, the estimate of the optimal software rejuvenation schedule based on Dohi et al.'s[5,6] algorithm asymptotically converges on the real optimal solution. Recently, Rinsaka and Dohi[7] proposed the non-parametric estimation algorithm based on the kernel density estimation[8,9,10,11] to improve the estimation accuracy of the optimal software rejuvenation schedule with a complete small sample data.

In many situations, however, it is difficult to collect all the failure time data, since software system may be shot down or rebooted before the system failure. Then, some observations may be censored or truncated from the right, referred to as right-censorship. Data of this type are called censored data. Reineke et al.[12,13] proposed the non-parametric estimation algorithm based on the Kaplan-Meier estimation[14] to obtain the optimal age replacement times from the censored sample of failure time data.

The aim of the present chapter is to improve the estimation accuracy of the optimal software rejuvenation schedules from the censored sample data. More precisely, we propose a statistical estimation algorithm based on the kernel density estimation under random censorship[16,15] to obtain the optimal software rejuvenation schedules which maximize the steady-state system availability. In simulation experiments, we check the effect of improvement in terms of the convergence speed to the real optimal solution in comparison with the conventional algorithm.

2. Semi-Markov Models

2.1. *Model 1*

First, we introduce the software rejuvenation model proposed by Dohi *et al.*[5,6] which is an extension of CTMC model by Huang *et al.*[4] The model based on semi-Markov process has following four states:

State 0: highly robust state (normal operation state)
State 1: failure probable state
State 2: failure state
State 3: software rejuvenation state.

Here, State 1 means that the memory leakage is over a threshold or the system lapses from the highly robust state into an unstable state. Let Z be the random time interval when the highly robust state changes to the failure probable state, having the common distribution function $\Pr\{Z \leq t\} = F_0(t)$ with finite mean μ_0 (> 0). Just after the state becomes the failure probable state, system failure may occur with positive probability. Without loss of generality, we assume that the random variable Z is observable during the system operation[4,17]. Let X denote the failure time from State 1, having the distribution function $\Pr\{X \leq t\} = F_f(t)$ with finite mean λ_f (> 0). If the system failure occurs before triggering a software rejuvenation, then the recovery operation starts immediately at that time. Otherwise the software rejuvenation starts. Let Y be the random repair time from the failure state, having the common distribution function $\Pr\{Y \leq t\} = F_a(t)$ with finite mean μ_a (> 0). Note that the software rejuvenation cycle is measured from the time instant just after the system enters State 1 from State 0. Denote the distribution function of the time to invoke the software rejuvenation and the distribution of the time to complete software rejuvenation by $F_r(t)$ and $F_c(t)$ (with mean μ_c (> 0)), respectively. After completing the repair or the rejuvenation, the software system becomes as good as new, and the software age is initiated at the beginning of the next highly robust state. Consequently, we define the time interval from the beginning of the system

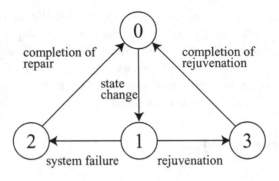

Figure 1. Semi-Markovian diagram of Model 1.

operation to the next one as one cycle, and the same cycle repeats again and again. It is noted that all the states in State 0 ∼ State 3 are regeneration points. The transition diagram for Model 1 is depicted in Fig. 1. If we consider the time to software rejuvenation time as a constant t_0, then it follows that

$$F_r(t) = U(t - t_0) = \begin{cases} 1 & \text{if } t \geq t_0 \\ 0 & \text{otherwise,} \end{cases} \qquad (1)$$

where $U(\cdot)$ is the unit step function. We call t_0 (≥ 0) the software rejuvenation schedule in this chapter. Hence, the underlying stochastic process is a semi-Markov process with four regeneration states. Note that under the assumption that the sojourn times in all states are exponentially distributed, this model is reduced to Huang et al.'s CTMS model[4]. Applying the standard technique of semi-Markov processes[5], the steady-state system availability for Model 1, $A_1(t_0)$ becomes:

$$A_1(t_0) = \Pr\{\text{software system is operative in the steady state}\}$$
$$= \frac{\mu_0 + \int_0^{t_0} \overline{F}_f(t)dt}{\mu_0 + \mu_a F_f(t_0) + \mu_c \overline{F}_f(t_0) + \int_0^{t_0} \overline{F}_f(t)dt}, \qquad (2)$$

where in general $\overline{F}(\cdot) = 1 - F(\cdot)$.

2.2. Model 2

The next model is a modification of Model 1[5]. In this model, we assume that the software system is not renewed even when the recovery operation is completed after the system failure. If we distinguish between the recovery operation and the software rejuvenation, then an additional rejuvenation

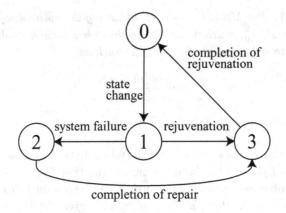

Figure 2. Semi-Markovian diagram of Model 2.

may be needed after the recovery operation. For example, restarting the system after repair might require some cleanup and resuming the process execution at the checkpoint. Figure 2 is the semi-Markov diagram for Model 2. In this model, the software rejuvenation is performed just after the completion of recovery as well as at the constant time t_0 after the failure probable state is entered, i.e., $\min\{Z + t_0, Z + X + Y\}$.

The system availability for Model 2, which is the probability that the software system is operating in the steady state, is given by

$$A_2(t_0) = \frac{\mu_0 + \int_0^{t_0} \overline{F}_f(t)dt}{\mu_0 + \mu_c + \mu_a F_f(t_0) + \int_0^{t_0} \overline{F}_f(t)dt}. \quad (3)$$

3. The TTT Concept

To derive the optimal software rejuvenation schedules on the graph, we define the scaled total time on test (TTT) transform[18] of the failure time distribution:

$$\phi(p) = \frac{1}{\lambda_f} \int_0^{F_f^{-1}(p)} \overline{F}_f(t)dt, \quad (4)$$

where

$$F_f^{-1}(p) = \inf\{t_0; F_f(t_0) \geq p\}, \quad 0 \leq p \leq 1. \quad (5)$$

It is well known[18] that $F_f(t)$ is IFR (DFR) if and only if $\phi(p)$ is concave (convex) on $p \in [0, 1]$. Dohi et al.[5] show the following result.

Theorem 3.1. *For Model i ($i = 1, 2$), obtaining the optimal software rejuvenation schedule, t_0^*, maximizing the steady-state system availability $A_i(t_0)$ is equivalent to obtaining p^* ($0 \leq p^* \leq 1$) such as*

$$\max_{0 \leq p \leq 1} \frac{\phi(p) + \alpha}{p + \eta_i}, \qquad (6)$$

where $\alpha = \mu_0/\lambda_f$, $\eta_1 = \mu_c/(\mu_a - \mu_c)$ and $\eta_2 = \mu_c/\mu_a$.

Theorem 3.1 can be obtained by transforming $A_i(t_0)$ to a function of p by means of $p = F_f(t)$. If the failure time distribution $F_f(t)$ is known, then the optimal software rejuvenation schedule can be obtained from Theorem 3.1 by $t_0^* = F_f^{-1}(p^*)$. Here, p^* ($0 \leq p^* \leq 1$) is given by the x coordinate value p^* for the point of the curve with the largest slope among the line pieces drawn from the point $(-\eta_i, -\alpha) \in (-\infty, 0) \times (-\infty, 0)$ to the curve $(p, \phi(p)) \in [0, 1] \times [0, 1]$ on a two-dimensional plane.

4. The Kaplan-Meier Estimator

Often in testing phase, as well as in operational situations, it is difficult to observe all the failure times. This is because software system may undergo shutdown or reboot before the system failure. Failure time data often include the data which does not fail during the testing and the operational phases. Such data are said to be right-censored. Let $X_1, X_2, \cdots, X_n$ denote the true survival times which are censored on the right by a sequence $U_1, U_2, \cdots, U_n$ which in general may be either constants or random variables.

The observed right-censored data are denoted by the pairs (W_j, Δ_j), $j = 1, \cdots, n$, where

$$W_j = \min\{X_j, U_j\}, \quad \Delta_j = \begin{cases} 1 & \text{if } X_j \leq U_j, \\ 0 & \text{if } X_j > U_j. \end{cases} \qquad (7)$$

Thus, it is known which observations are times of failure and which ones are censored or loss times. In this chapter, we assume that $U_1, \cdots, U_n$ constitute a random sample from a distribution G (which is usually unknown) and are independent of $X_1, \cdots, X_n$. That is, (W_j, Δ_j), $j = 1, 2, \cdots, n$, is called a randomly right-censored sample.

Based on the censored sample (W_j, Δ_j), $j = 1, \cdots, n$, a popular estimator of the survival probability is the Kaplan-Meier estimator[14] as the non-parametric maximum likelihood estimator of $R(t) = 1 - F_f(t)$. Let

$(W_{(j)}, \Delta_{(j)})$, $j = 1, \cdots, n$, denote the ordered W_j's along with the corresponding Δ_j's. The Kaplan-Meier estimator of R is defined by

$$\hat{R}_{\text{KME}}(t) = \begin{cases} 1, & 0 \le t < W_{(1)}, \\ \prod_{j=1}^{k-1} \left(\frac{n-j}{n-j+1}\right)^{\Delta_{(j)}}, & W_{(k-1)} \le t < W_{(k)}, \ k = 2, \cdots, n, \\ 0, & t \ge W_{(n)}. \end{cases}$$
(8)

Let s_j denote the jump of $\hat{R}_{\text{KME}}$ at $W_{(j)}$, that is,

$$s_j = \hat{R}_{\text{KME}}(W_{(j-1)}) - \hat{R}_{\text{KME}}(W_{(j)}), \quad j = 1, \cdots, n. \tag{9}$$

Note that $s_j = 0$ if and only if $\Delta_j = 0$, $j \le n$, that is, if $W_{(j)}$ is a censored observation.

Let $\chi_1, \chi_2, \cdots, \chi_m$ denote the observed failure times and let $\chi_{(1)} \le \chi_{(2)} \le \cdots \le \chi_{(m)}$ denote the order statistics of the χ_j, where m ($\le n$) is the number of observed (uncensored) failures. For randomly censored data, the TTT-plot can be constructed using the Kaplan-Meier estimator by letting $p_{(j)} = 1 - \hat{R}_{\text{KME}}(\chi_{(j)})$, $j = 1, 2, \cdots, m$, for the ordered failure time j and by estimating the TTT-transform with

$$\hat{H}_{\text{KME}}^{-1}(p_{(j)}) = \int_0^{\chi_{(j)}} \hat{R}_{\text{KME}}(t) dt$$

$$= \sum_{k=1}^{j} \left(\chi_{(k)} - \chi_{(k-1)}\right) \hat{R}_{\text{KME}}(\chi_{(k-1)}),$$

$$j = 1, 2, \cdots, m; \quad \chi_{(0)} = 0. \tag{10}$$

The TTT-plot is obtained by plotting the coordinates

$$\left\{ p_{(j)}, \frac{\hat{H}_{\text{KME}}^{-1}(p_{(j)})}{\hat{H}_{\text{KME}}^{-1}(p_{(m)})} \right\}, \quad j = 1, 2, \cdots, m. \tag{11}$$

By connecting the points in a staircase pattern, the scaled TTT plot is obtained. Since the estimate in Eq.(11) is a non-parametric estimate of $(p, \phi(p))$, $p \in [0, 1]$, the following theorem on the optimal software rejuvenation schedule is obtained by direct application of the result in Theorem 3.1.

Theorem 4.1. *It is assumed that the randomly censored failure time data (w_j, δ_j), $j = 1, \cdots, n$ are observed. For Model i ($i = 1, 2$), the non-parametric estimate $\hat{t}_0^*$ of an optimal software rejuvenation schedule maximizing the steady-state system availability is given by $\chi_{(j^*)}$ satisfying the* following:

$$j^* = \left\{ j \,\Big|\, \max_{0 \le j \le m} \frac{\hat{H}_{\text{KME}}^{-1}(p_{(j)})/\hat{H}_{\text{KME}}^{-1}(p_{(m)}) + \mu_0/\hat{H}_{\text{KME}}^{-1}(p_{(m)})}{p_{(j)} + \eta_i} \right\}. \quad (12)$$

5. The Kernel Density Estimation

In this section, we propose the kernel density estimation to obtain the optimal software rejuvenation schedule from the censored sample data with higher accuracy. Suppose that the true failure times $X_1, \cdots, X_n$ are the non-negative independent identically distributed random variables with common unknown distribution function F_f and the density function f_f. Again, we assume that the right-censored data (w_j, δ_j), $j = 1, \cdots, n$ can be observed. Then, we define the kernel density estimator[15,16] by

$$\hat{f}_{f\text{KDE}}(y) = \frac{1}{h} \sum_{j=1}^{n} s_j K\left(\frac{y - w_j}{h}\right), \quad (13)$$

where s_j is given by Eq.(9). The parameter h (> 0) is the window width called the *smoothing parameter* or *bandwidth*, and the function $K(\cdot)$ is called the *kernel function* which satisfies the following condition:

$$\int_{-\infty}^{\infty} K(t)dt = 1, \quad \int_{-\infty}^{\infty} tK(t)dt = 0, \quad 0 < \int_{-\infty}^{\infty} t^2 K(t)dt < \infty. \quad (14)$$

Usually, but not always, the function $K(\cdot)$ is selected as a symmetric probability density function. In this chapter, we apply the following Gaussian kernel function:

$$K(t) = \frac{1}{\sqrt{2\pi}} e^{-(1/2)t^2} \quad (15)$$

to estimate the density function of the system failure time. The main reason to use it is its tractability and convergence property.

Now, define the scaled total time on test transform of the estimator $\hat{F}_{f\text{KED}}(t) = 1 - \hat{R}_{\text{KED}}(t) = \int_0^t \hat{f}_{f\text{KED}}(s)ds$ of failure time distribution by

$$\phi_{\text{KDE}}(p) = \frac{1}{\hat{\lambda}_{n\text{KDE}}} \int_0^{\hat{F}_{f\text{KDE}}^{-1}(p)} \hat{R}_{\text{KDE}}(t)dt, \quad (16)$$

where $\hat{\lambda}_{n\text{KDE}}$ is the estimate of mean time to failure and can be estimated as

$$\hat{\lambda}_{n\text{KDE}} = \int_0^{\infty} \hat{R}_{\text{KDE}}(t)dt. \quad (17)$$

Table 1. Censoring parameters and statistically expected proportion of censoring.

q	0.1	0.2	0.3	0.4	0.5	0.6	0.7	0.8	0.9
ν	1326.4	616.0	378.0	258.0	184.9	135.0	97.9	68.1	41.4

The following theorem on the optimal software rejuvenation schedule is obtained by direct application of the result in Theorem 3.1.

Theorem 5.1. *It is assumed that the randomly censored failure time data (w_j, δ_j), $j = 1, \cdots, n$ are observed. For Model i ($i = 1, 2$), the non-parametric estimate $\hat{t}_0^*$ of an optimal software rejuvenation schedule maximizing the steady-state system availability is given by $\hat{t}_0^* = \hat{F}_{fKDE}^{-1}(p^*)$ satisfying the following maximization problem:*

$$\max_{0 \leq p \leq 1} \frac{\phi_{KDE}(p) + \hat{\lambda}_{nKDE}}{p + \eta_i}. \tag{18}$$

6. Simulation Experiments

Of our next interest is the investigation of asymptotic properties and convergence speed of estimators proposed in this chapter. Suppose that the failure time obeys the Weibull distribution:

$$F_f(t) = 1 - e^{-(t/\theta)^\gamma}, \quad \gamma > 0, \ \theta > 0. \tag{19}$$

We assume that the random censoring time distribution is the exponential distribution:

$$G(t) = 1 - e^{-t/\nu}, \quad \nu > 0. \tag{20}$$

The statistically expected proportion of censoring, q, is given by

$$q = \int_0^\infty \overline{F}_f(t) dG(t). \tag{21}$$

In the following simulation experiments, the Weibull shape and scale parameters are fixed as $\gamma = 2.0$ and $\theta = 160.0$. Table 1 presents the censoring parameters and corresponding statistically expected proportion of censoring. The other parameters are fixed as $\mu_0 = 24.0$, $\mu_a = 1.0$ and $\mu_c = 1/3$. In this situation, if we can know the probability distribution $F_f(t)$ completely, then the optimal software rejuvenation schedule and its associated maximum system availability for Model 1 can be calculated as $t_0^* = 93.67$ and $A_1(t_0^*) = 0.99515$, respectively. For Model 2, we obtain $t_0^* = 72.34$ and $A_2(t_0^*) = 0.99438$.

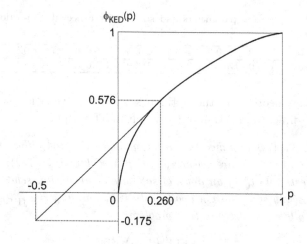

Figure 3. Estimation of the optimal software rejuvenation schedule based on the kernel density estimation (Model 1).

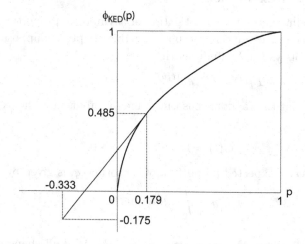

Figure 4. Estimation of the optimal software rejuvenation schedule based on the kernel density estimation (Model 2).

Let us consider the estimation of an optimal software rejuvenation schedule when the randomly right-censored failure time data are already observed. It is assumed that the observed data consist of 30 pseudo random numbers generated from the Weibull failure time distribution in Eq.(19) and the exponential censoring time distribution in Eq.(20), where

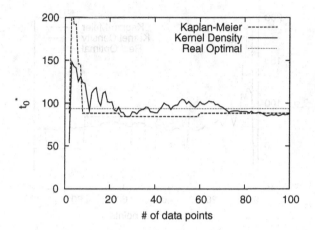

Figure 5. Asymptotic behavior of estimate of the optimal preventive rejuvenation schedule form Model 1.

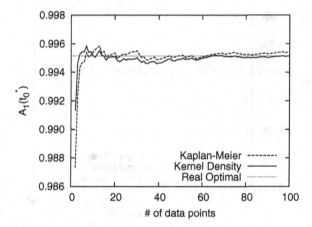

Figure 6. Asymptotic behavior of estimate of the maximum system availability for Model 1.

the s-expected proportion of censoring is $q = 0.2$. In Figs. 3 and 4, we illustrate estimation examples of the optimal software rejuvenation schedules based on the kernel density estimation from 30 observed right-censored data for Model 1 and Model 2, respectively. In Fig. 3, the point providing the steepest slope among the line segments drawn from $(-\eta_1, \hat{\lambda}_{n\text{KDE}}) = (-0.5, -0.175)$ to the scaled TTT transform $\phi_{\text{KDE}}(p)$ is $p^* = 0.260$. Hence we estimate the optimal software rejuvenation schedule and the maximum system availability for Model 1 as $\hat{t}_0^* = 87.12$ and

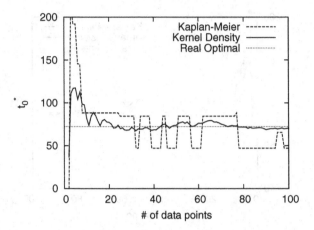

Figure 7. Asymptotic behavior of estimate of the optimal preventive rejuvenation schedule form Model 2.

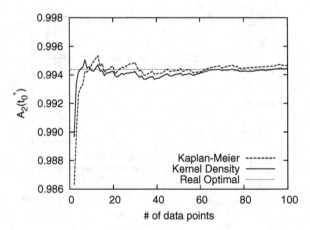

Figure 8. Asymptotic behavior of estimate of the maximum system availability for Model 2.

$A_1(\hat{t}_0^*) = 0.99511$. In Fig. 4, the point providing the steepest slope among line segments drawn from $(-\eta_2, \hat{\lambda}_{nKDE}) = (-0.333, -0.175)$ to $\phi_{KDE}(p)$ is $p^* = 0.179$. The optimal software rejuvenation schedule and the maximum system availability for Model 2 can be estimated as $\hat{t}_0^* = 71.20$ and $A_2(\hat{t}_0^*) = 0.99438$.

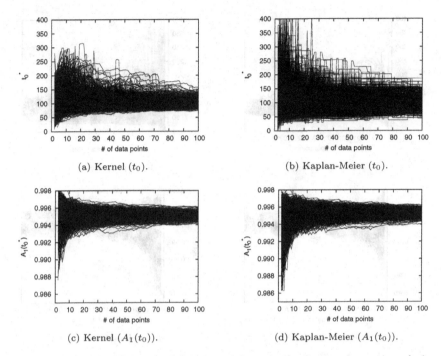

Figure 9. Asymptotic behavior of estimates of the optimal software rejuvenation schedule and the maximum system availability for 1,000 simulation runs (Model 1).

Next, we investigate the asymptotic behavior of the kernel density estimates and compare it with the existing Kaplan-Meier estimation approach. To do it, the Monte Carlo simulations are carried out with pseudo random numbers based on the Eqs.(19) and (20). Figures 5 through 8 show the asymptotic behavior of estimates of the optimal software rejuvenation schedule and the maximum system availability. Figures 9 and 10 plot the estimates of the optimal software rejuvenation schedule and the maximum system availability for 1,000 simulation runs. From these figures, it is found that the results converge to the real optimal solutions when the number of failure time data is close to 30. The range of the estimates of the kernel method is small as compared with that of the Kaplan-Meier method.

Finally, we examine the convergence speed and accuracy of the kernel method for the different censoring time distribution. In Figs. 11 through 14, we calculate the relative absolute error averages, $RAEA_{t_0}$ and $RAEA_{A_i}$, of estimates of the optimal software rejuvenation schedule and maximum

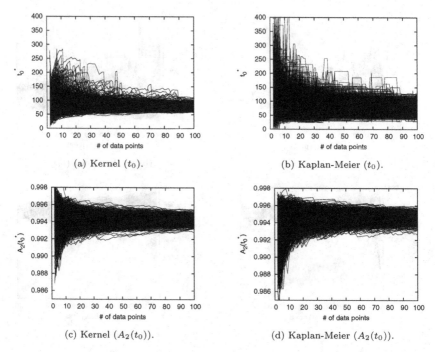

Figure 10. Asymptotic behavior of estimates of the optimal software rejuvenation schedule and the maximum system availability for 1,000 simulation runs (Model 2).

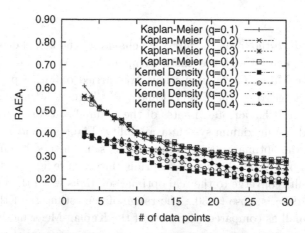

Figure 11. Relative absolute error average of estimates of the optimal software rejuvenation schedule (Model 1).

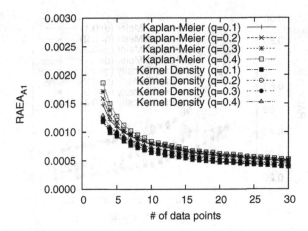

Figure 12. Relative absolute error average of estimates of the maximum system availability (Model 1).

system availability, where

$$\text{RAEA}_{t_0} = \frac{1}{mt_0^*} \sum_{j=1}^{m} \left| \hat{t}_{0_j}^* - t_0^* \right| \qquad (22)$$

and

$$\text{RAEA}_{A_i} = \frac{1}{mA_i(t_0^*)} \sum_{j=1}^{m} \left| A_i(\hat{t}_{0_j}^*) - A_i(t_0^*) \right|, \quad i = 1, 2. \qquad (23)$$

In this experiment, $m = 1,000$ simulation runs are executed with $q = 0.1$, 0.2, 0.3 and 0.4, where $\hat{t}_{0_j}^*$ is the optimal software rejuvenation schedule estimated in the j-th simulation run. From Figs. 11 through 14, we can see that the relative absolute error average can be reduced as the s-expected proportion q of censoring becomes small. For small sample data, we can observe that the convergence speed of the optimal software rejuvenation schedule estimated by the kernel density estimation is faster than by the Kaplan-Meier method. From these results, we conclude that the statistical algorithm based on the kernel density estimation can be recommended to estimate the optimal software rejuvenation schedule under censoring.

7. Concluding Remarks

In this chapter, we have considered the optimal software rejuvenation schedule and developed the statistical estimation algorithms from the randomly censored failure time data. The non-parametric estimation algorithm based

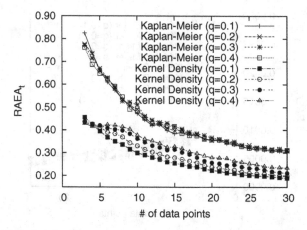

Figure 13. Relative absolute error average of estimates of the optimal software rejuvenation schedule (Model 2).

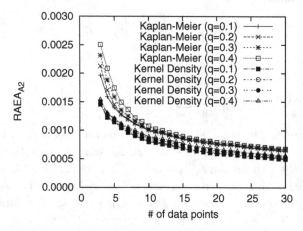

Figure 14. Relative absolute error average of estimates of the maximum system availability (Model 2).

on the kernel density estimation has been proposed to improve the estimation accuracy for the randomly censored small sample of failure time data. Throughout simulation experiments, it has been shown that the proposed algorithm based on the kernel density estimation had higher estimation accuracy than the Kaplan-Meier method, and faster convergence speed to the theoretical optimal software rejuvenation schedule.

Acknowledgments

The present research was partially supported by the Ministry of Education, Culture, Sport, Science and Technology, Grant-in-Aid for Young Scientists (B); Grant No. 18710145 (2006–2007) and Scientific Research (C); Grant No. 19510148 (2007–2008).

References

1. E. Adams, Optimizing preventive service of the software products, *IBM J. Research & Development*, **28** (1), pp. 2–14, (1984).
2. V. Castelli, R. E. Harper, P. Heidelberger, S. W. Hunter, K. S. Trivedi, V. Vaidyanathan and W. P. Zeggert, Proactive management of software aging, *IBM J. Research & Development*, **45** (2), pp. 311–332, (2001).
3. J. Gray and D. P. Siewiorek, High-availability computer systems, *IEEE Comput.*, **24** (9), pp. 39–48, (1991).
4. Y. Huang, C. Kintala, N. Kolettis and N. D. Funton, Software rejuvenation: analysis, module and applications, *Proc. 25th IEEE Int'l Symp. Fault Tolerant Computing*, pp. 381–390, IEEE Computer Society Press, Los Alamitos, CA, (1995).
5. T. Dohi, K. Goševa-Popstojanova and K. S. Trivedi, Statistical nonparametric algorithms to estimate the optimal software rejuvenation schedule, *Proc. 2000 Pacific Rim Int'l Symp. on Dependable Computing*, pp. 77–84, IEEE Computer Society Press, Los Alamitos, CA, (2000).
6. T. Dohi, K. Goševa-Popstojanova and K. S. Trivedi, Analysis of software cost models with rejuvenation, *Proc. 5th IEEE Int'l Symp. High Assurance Systems Engineering*, pp. 25–34, IEEE Computer Society Press, Los Alamitos, CA, (2000).
7. K. Rinsaka and T. Dohi, Estimating the optimal software rejuvenation schedule with small sample data, *Proceedings of 2006 Asian International Workshop on Advanced Reliability Modeling (AIWARM 2006)*, pp. 443–450, World Scientific, Singapore, (2006).
8. T. Cacoullos, Estimation of a multivariate density, *Annals of the Institute of Statistical Mathematics*, **18** (2), pp. 178–189, (1966).
9. E. Parzen, On the estimation of a probability density function and the mode, *Annals of Mathematical Statistics*, **33**, pp. 1065–1076, (1962).
10. M. Rosenblatt, Remarks on some nonparametric estimates of a density function, *Annals of Mathematical Statistics*, **27**, pp. 832–837, (1956).
11. B. W. Silverman, *Density Estimation for Statistics and Data Analysis*, Chapman and Hall, London, (1986).
12. D. M. Reinke, E. A. Pohl and W. P. Murdock Jr, Survival analysis and maintenance policies for a series system, with highly censored data, *Proc. Annual Reliability and Maintainability Symposium*, pp. 182–188, (1998).
13. D. M. Reinke, E. A. Pohl and W. P. Murdock Jr, Maintenance-policy cost-analysis for a series system with highly-censored data, *IEEE Trans. Reliab.*, **R-48** (4), pp. 413–420, (1999).

14. E. L. Kaplan and P. Meier, Nonparametric estimation from incomplete observations, *J. Amer. Statist. Assoc.*, **53**, pp. 457–481, (1958).
15. D. T. McNichols and W. J. Padgett, Kernel density estimation under random censorship, *Statistics Tech. Rep.*, **74**, University of South Carolina, (1981).
16. W. J. Padgett, Nonparametric estimation of density and hazard rate functions when samples are censored, in *Handbook of Statistics*, **7** (eds. P.R. Krishnaiah and C.R. Rao), pp. 313–331, North-Holland, New York, (1988).
17. S. Garg, Y. Huang, C. Kintala and K. S. Trivedi, Time and load based software rejuvenation: policy, evaluation and optimality, *Proc. 1st Fault-Tolerant Symp.*, pp. 22–25, (1995).
18. R. E. Barlow and R. Campo, Total time on test processes and applications to failure data, *Reliability and Fault Tree Analysis*, (eds. R. E. Barlow, J. Fussell and N. D. Singpurwalla), pp. 451–481, SIAM, Philadelphia, (1975).

OPTIMAL BACKUP INTERVAL OF A DATABASE SYSTEM USING A CONTINUOUS DAMAGE MODEL

SYOUJI NAKAMURA*

Department of Human Life and Information,
Kinjo Gakuin University
1723 Omori 2-chome, Moriyama-ku, Nagoya 463-8521, Japan
snakam@kinjo-u.ac.jp

TOSHIO NAKAGAWA*

Department of Management and Information Systems,
Aichi Institute of Technology
1247 Yachigusa, Yakusa-cho, Toyota 470-0392, Japan
toshi-nakagawa@aitech.ac.jp

HITOSHI KONDO

Faculty of Economics,
Nanzan University
118 Yamazato-cho, Showa-ku, Nagoya 466-8673, Japan
hitoshi@nanzan-u.ac.jp

Suppose that a database is updated and an amount of updated files accumulates additively. Some media failures occur at random. A full backup is made when the total updated files exceed a threshold level. To lessen the overhead of backup processing, the operation of an incremental backup with small overhead is adopted between the full backup. The mean time to incremental or full backup and the expected cost of these schemes are derived, using the theory of cumulative processes. Further, optimal numbers of incremental backup which minimize the expected costs are analytically discussed. A numerical example is finally given.

---

*This work is supported in part by the Grant-in-Aid for Scientific Research (C), Grant No. 19530296 (2007-2008) from the Ministry of Education, Culture, Sports, Science, and Technology of Japan.

1. Introduction

Databases in a computer system are frequently updated by adding or deleting data, and are stored in other secondary media. However, data files in secondary media are sometimes broken by several errors due to noises, human errors and hardware faults. In this case, we have to reconstruct the same files from the beginning. The most simple and dependable method to ensure the safety of data would be always to make the backup copies of all files in other places, and to take out them if files in the original secondary media are broken. This is called the full backup [1]. But, this method would take hours and costs when files become large.

To lessen the overhead of backup processing, the operation of an incremental backup with small overhead is adopted between the full backup. The incremental backup exports only updated files which have changed or are new since the last full backup, when the total updated files do not exceed a threshold level K. Optimal policies for backup schemes were discussed analytically in [1-3].

We can formulate the above backup model of a database by transforming *shock* into *update* and *damage* into *updated files* in the cumulative damage model [4,5]. This paper applies a continuous wear process [6,7] to this model: A database is updated continuously and its total updated files $Z(t)$ increase linearly with time t, *i.e.*, $Z(t) = At$. The full backup is done when the total files exceeds a threshold level K.

Suppose that the backup can be done only at periodic times nT ($n = 1, 2, \ldots$) for a specified T. Then, the full backup is always done when the total files exceed K and the incremental backup is done at time NT when they do not exceed K. We can make the recovery of a database by importing files of the last incremental or full backup, if some errors have occurred in a storage media.

In this paper, we consider two particular cases where K and A are random variables. Then, introducing the costs of two backups and using the theory of cumulative processes [4], we obtain the expected cost rates. Further, we discuss analytically optimal numbers N^* of incremental backup that minimize the expected cost rates for each case. It is shown that optimal N^* are given by unique solutions of some equations. As a numerical example, an optimal number N^* is computed when K is distributed normally.

2. Expected Cost

Suppose that a database is updated continuously and its total update files $Z(t)$ increase linearly with its operating time, *i.e.*, $Z(t) = At$. It is assumed that the total files are checked only at planned times nT $(n = 1, 2, \dots)$ for a specified $T > 0$. If the total files exceed a threshold level K during the interval $((n-1)T, nT]$, the full backup is done at time nT and the database returns to an initial state, *i.e.*, $Z(nT+0) = 0$. An event that the total files exceed K occurs independently during each interval. Further, the incremental backup is done at a planned time NT before the total files exceed K.

We consider the following two particular cases: First, a threshold level K is a random variable with a general distribution $G(x) \equiv \Pr\{K \leq x\}$ and a parameter $A(>0)$ is constant (Figure 1). Then, the probability that the full backup is done at time nT is $\Pr\{AnT \geq K\} = G(AnT)$. Thus, the probability that the full backup is done until time NT is

$$\sum_{n=1}^{N} G(AnT) \prod_{j=0}^{n-1} \overline{G}(AjT), \tag{1}$$

and the probability that the incremental backup is done at time NT is

$$\prod_{n=1}^{N} \overline{G}(AnT), \tag{2}$$

where $\overline{\Phi}(x) \equiv 1 - \Phi(x)$ for any function $\Phi(x)$. Further, the mean time to the backup is

$$\sum_{n=1}^{N}(nT)G(AnT)\prod_{j=0}^{n-1}\overline{G}(AjT) + (NT)\prod_{n=1}^{N}\overline{G}(AnT) = T\sum_{n=0}^{N-1}\left[\prod_{j=0}^{n}\overline{G}(AjT)\right], \tag{3}$$

and hence, the mean time to the full backup is

$$l_1 = T\sum_{n=0}^{\infty}\left[\prod_{j=0}^{n}\overline{G}(AjT)\right]. \tag{4}$$

Let c_K be the full backup cost and $c_N(< c_K)$ be the incremental backup cost. Then, from (1),(2) and (3), the expected cost rate is [8,9]

$$C_1(N) = \frac{c_K - (c_K - c_N) \prod_{n=1}^{N} \overline{G}(AnT)}{T \sum_{n=0}^{N-1} \left[\prod_{j=0}^{n} \overline{G}(AjT) \right]} \quad (N = 1, 2, \ldots). \quad (5)$$

Next, a parameter A is a random variable with a general distribution $L(x) \equiv \Pr\{A \leq x\}$ and a threshold level $K(> 0)$ is constant (Figure 1). Then, the probability that the full backup is done at time nT is $\Pr\{AnT \geq K\} = \overline{L}(K/(nT))$. Thus, the probability that the full backup is done until time NT is

$$\overline{L}(K/T) + \sum_{n=1}^{N} \overline{L}(K/((n+1)T)) \prod_{j=1}^{n} L(K/(jT)), \quad (6)$$

and the probability that the increment backup is done at time NT is

$$\prod_{n=1}^{N} L(K/(nT)). \quad (7)$$

It is noted that (1)+(2)=1. Further, the mean time to the backup is

$$T\overline{L}(K/T) + \sum_{n=1}^{N} [(n+1)T] \overline{L}(K/((n+1)T)) \prod_{j=1}^{n} L(K/(jT))$$
$$+ (NT) \prod_{n=1}^{N} L(K/(nT)) = T \left\{ 1 + \sum_{n=1}^{N-1} \left[\prod_{j=1}^{n} L(K/(jT)) \right] \right\}, \quad (8)$$

and hence, the mean time to the full backup is

$$l_2 = T \left\{ 1 + \sum_{n=1}^{\infty} \left[\prod_{j=1}^{n} L(K/(jT)) \right] \right\}. \quad (9)$$

Therefore, the expected cost rate is, from (7) and (8),

$$C_2(N) = \frac{c_K - (c_K - c_N)\prod_{n=1}^{N} L[K/(nT)]}{T\left\{1 + \sum_{n=1}^{N-1}\left[\prod_{j=1}^{n} L(K/(jT))\right]\right\}} \quad (N = 1, 2, \ldots), \quad (10)$$

where $\sum_{n=1}^{0} \equiv 0$.

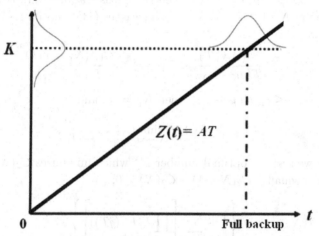

Figure 1. Full backup interval

3. Optimal Policy

If both K and A are constant, then we may do the full backup immediately before $t = K/A$. When K is a random variable, we seek an optimal number N_1^* which minimizes $C_1(N)$ in (5). From the inequality $C_1(N+1) - C_1(N) \geq 0$,

$$G(A(N+1)T)\sum_{n=0}^{N-1}\left[\prod_{j=0}^{n}\overline{G}(AjT)\right] + \prod_{n=1}^{N}\overline{G}(AnT) \geq \frac{c_K}{c_K - c_N}$$

$$(N = 1, 2, \ldots). \quad (11)$$

Letting $Q_1(N)$ be the left-hand side of (11),

$$Q_1(\infty) \equiv \lim_{N \to \infty} Q_1(N) = \sum_{n=0}^{\infty}\left[\prod_{j=0}^{n}\overline{G}(AjT)\right] = \frac{l_1}{T},$$

$$Q_1(N+1) - Q_1(N) =$$

$$[G(A(N+2)T) - G(A(N+1)T)]\sum_{n=0}^{N}\left[\prod_{j=0}^{n}\overline{G}(AjT)\right] > 0.$$

Thus, $Q_1(N)$ is strictly increasing to l_1/T, and hence, we have the optimal backup policy:

(i) If $l_1/T > c_K/(c_K - c_N)$, then there exists a finite and unique minimum $N_1^*(1 \le N_1^* < \infty)$ which satisfies (11), and its resulting cost rate is

$$\frac{G(AN_1^*T)}{T(c_K - c_N)} < C_1(N_1^*) \le \frac{G(A(N_1^*+1)T)}{T(c_K - c_N)}. \tag{12}$$

(ii) If $l_1/T \le c_K/(c_K - c_N)$, then $N_1^* = \infty$, and

$$C_1(\infty) \equiv \lim_{N \to \infty} C_1(\infty) = \frac{c_K}{l_1}. \tag{13}$$

Next, we seek an optimal number N_2^* which minimizes $C_2(N)$ in (10). From the inequality $C_2(N+1) - C_2(N) \ge 0$,

$$\overline{L}(K/((N+1)T))\left\{1 + \sum_{n=1}^{N-1}\left[\prod_{j=1}^{n}L(K/(jT))\right]\right\}$$

$$+ \prod_{n=1}^{N}L(K/(nT)) \ge \frac{c_K}{c_K - c_N} \quad (N = 1, 2, \ldots). \tag{14}$$

Letting $Q_2(N)$ be the left-hand side of (14),

$$Q_2(\infty) \equiv \lim_{N \to \infty} Q_2(N) = 1 + \sum_{n=1}^{\infty}\left[\prod_{j=1}^{n}L(K/(jT))\right] = \frac{l_2}{T},$$

$$Q_2(N+1) - Q_2(N) =$$

$$[\overline{L}(K/((N+2)T)) - \overline{L}(K/((N+1)T))]\left\{1 + \sum_{n=1}^{N}\left[\prod_{j=1}^{n}L(K/(jT))\right]\right\} > 0.$$

Thus, $Q_2(N)$ is strictly increasing to l_2/T, and hence, replacing l_1 with l_2, we have the similar optimal policy to the previous one.

4. Numerical Example

Suppose that a threshold level K at which the full backup is done is normally distributed with mean k and standard deviation σ, and $AT = 1$, i.e.,

$$\overline{G}(j) = \frac{1}{\sqrt{2\pi}\sigma} \int_j^\infty \exp[-(x-k)^2/(2\sigma^2)] \mathrm{d}x \quad (j = 0, 1, 2, \dots). \quad (15)$$

Table 1 presents the optimal incremental backup number N_1^* and the mean time l_1 to full backup execution for $k = 10, 20, 50$ and $\sigma = 1, 2, 5, 10$ when $c_K/c_N = 5$.

Another single method of such backup systems is to balance the cost of full backup against incremental backup; i.e., $c_K \times (1) \geq c_N \times (2)$. In this case,

$$\prod_{j=1}^N \overline{G}(j) \leq \frac{c_K}{c_K + c_N}, \quad (16)$$

and a minimum $\widetilde{N}$ to satisfy it is also presented in Table 1.

For example, when the mean threshold level of updated files is $k = 20$ and $\sigma = 5$, the optimal incremental backup interval is $N_1^* = 8$, and the resulting minimum cost is $C_1(N_1^*) = 15.99$. Similarly, when $k = 50$ and $\sigma = 5$, the optimal interval is $N_1^* = 36$, and the resulting cost is $C_1(N_1^*) = 45.99$. From the comparison of these two examples, when the ratio of the values of each k is at 2.5 times, the ratio of the value N_1^* is 4.5 times. That is, when the threshold level grows, the full backup interval grows more than the ratio of threshold. When σ becomes large, N_1^* becomes small. It is shown that the incremental backup should be executed as early as possible when the change of the threshold level is large. The values of $\widetilde{N}$ in Table 1 give a good approximation as the upper bound of N_1^* for small σ.

We set usually an on-line processing operation for a day to a backup interval in an actual backup method. When an on-line processing of a database for every day is ended, an incremental backup can be executed. It has been known in actual applications of such processing that the mean and standard deviation of a threshold level k is approximately $k = 20$ and $\sigma = 5$. Therefore, as most backup methods, an incremental backup interval is about one week and is executed on Saturday or Sunday. Such backup interval should be decided by considering some fault occurrences, the possibility of its operation after the end of an on-line processing on weekdays, and so on. The backup scheme derived in this model and examples would be applied to actual database systems by roughly estimating backup operation costs.

Table 1. Comparative table of optimal number N_1^*, its approximate value $\tilde{N}$ and mean time l_1 to full backup when $AT = 1$ and $c_K/c_N = 5$.

σ	$k = 10$			$k = 20$			$k = 50$		
	N_1^*	$\tilde{N}$	l_1	N_1^*	$\tilde{N}$	l_1	N_1^*	$\tilde{N}$	l_1
1	7	9	10.27	17	19	20.27	47	49	50.27
2	5	8	9.51	15	18	19.51	44	48	49.51
5	2	4	6.40	8	13	15.99	36	43	45.99
10	1	1	3.92	3	5	9.55	23	33	38.38

5. Conclusions

We have analyzed the backup model of a database system by using the cumulative damage model: We have considered two cases where a threshold level at which the full backup is done is random and the total updated files increase with time t linearly and randomly. The expected cost rates for two cases have been obtained and the optimal incremental backup intervals which minimize them have been discussed analytically.

Further, we show that this would be applied to the backup of secondary storage files in a database system. Thus, by estimating backup costs and and amount of updated files from actual data and by modifying some suppositions, we could practically determine a planned time of an incremental backup. These formulations and results would be applied to other management policies for computer systems.

References

1. Qian C.H., Nakamura S. and Nakagawa T.(1999) Cumulative damage model with two kinds of shocks and its application to the backup policy, *J. of Operations Research Soc. of Japan*, *42*, 501–511.
2. Qian C.H., Pan Y. and Nakagawa T.(2002) Optimal policies for a database system with two backup schemes, *RAIRO Operations Research*, *36*, 227–235.
3. Nakamura S., Qian C.H., Fukumoto S. and Nakagawa T.(2003) Optimal backup policy for a database system with incremental and full backups, *Mathematical and Computer Modeling*, *38*, 1373–1379.
4. Cox, D.R. (1962) *Renewal Theory*, Methuen, London.

5. Nakagawa, T. (2007) *Shock and Damage Models in Reliability Theory*, Springer, London.
6. Reynolds D.S. and Savage I.R.(1971) Random wear models in reliability theory, *Advanced Applied Probability, 3*, 229–248.
7. Lemoine A.J. and Wenocur M.L.(1985) On failure modeling, *Naval Research Logistics Quarterly, 32*, 497–508.
8. Barlow R.E. and Proschan F. (1965) *Mathematical Theory of Reliability*, John Wiley & Sons, New York.
9. Nakagawa T.(2005) *Maintenance Theory of Reliability*, Springer, London.

OPERATIONAL SOFTWARE PERFORMANCE EVALUATION BASED ON THE NUMBER OF DEBUGGINGS WITH TWO KINDS OF RESTORATION SCENARIOS

K. TOKUNO AND S. YAMADA

Department of Social Systems Engineering
Faculty of Engineering, Tottori University
4-101, Koyama, Tottori-shi 680-8552, Japan
E-mail: toku@sse.tottori-u.ac.jp, yamada@sse.tottori-u.ac.jp

We develop the performance evaluation method for the multi-task software system for operational use. In particular, we consider two kinds of restoration scenarios; one involves the debugging activity and the other does not involve the debugging activity. The software reliability growth process and the upward tendency of difficulty in debugging in the dynamic environment is described by the Markovian software availability model. We assume that the cumulative number of tasks arriving at the system follows the homogeneous Poisson process. Then we can formulate the distribution of the number of tasks whose processes can be complete within a prespecified processing time limit with the infinite-server queueing model. From the model, several quantities for software performance measurement considering the real-time property can be derived. These quantities are given as the functions of time and the number of debuggings. Finally, we present several numerical examples of the quantities to analyze the relationship between the software reliability/ restoration characteristics and the system performance measurement.

1. Introduction

Today the engineering system of the service reliability engineering has a growing attention;[1,2,3,4] this aims at the establishment of the evaluation methods for the quality of service created by the use of the artificial industrial products as well as the inherent quality of the products. Considering the software systems are just the industrial products to provide the services for the users, especially in computer network systems, it is meaningful to discuss the performance evaluation methods for software systems oriented to the service reliability engineering. Recently, the consortium of the Service Availability Forum[5] has been created to develop the computing framework and the interface between hardware and software systems with high service availability.

The studies on performance evaluation considering reliability for computing systems have much been discussed from the viewpoint of the hardware configuration.[6,7] On the other hand, from the viewpoint of software system, the discussions on inherent quality/reliability evaluation such as the estimation of the residual fault content and the prediction of software failure time have much been conducted,[8,9] while there exist few studies on the reliability-related performance evaluation. Kimura et al.[10,11] have discussed the evaluation methods of the real-time property for the N-version programming and the recovery block software systems; these are well-known as the methodologies of the fault-tolerant software systems. However, Kimura's studies have just applied the framework for the analysis from the aspect of the hardware configuration to the fault-tolerant software systems and have not included the characteristics particular to software systems such as the reliability growth process and the upward tendency of difficulty in debugging.

In this paper, we discuss the performance evaluation method of the software systems considering the real-time property; this is the different approach from Kimura's studies. The real-time property is defined as the attribute that the system can complete the task within the stipulated response time limit.[12,13] We assume that the software system can process the plural tasks simultaneously. Furthermore, we consider the operation-oriented restoration-action scenarios in this modeling. Debugging activities corresponding to system downs occurring during the operation phase are not always performed since protracting an inoperable time may much affect the customers. This is a different policy from the testing phase in which debugging activities are always performed whenever a software failure occurs. Since the real-time property is one of the customer-oriented ones, we need to reflect on the operational situation in software performance/availability modeling. Here we consider two kinds of restoration actions during the operation phase; one involves debugging and the other does not involve debugging. Tokuno and Yamada[14] have proposed the Markovian software availability model considering the above operational restoration actions. We use this model to describe the time-dependent behavior of the software system itself alternating between up and down states. The stochastic behavior of the number of tasks whose processes can be complete within the prespecified processing time limit is modeled with the infinite sever queueing model.[15]

The organization of the rest of the paper is shown as follows. Section 2 states the software availability model used in the paper. Section 3 describes the stochastic processes of the numbers of tasks whose processes are complete within the prespecified processing time limit out of the tasks arriving up to a given time point. Section 4 derives several software performance measures considering the real-time property. These measures are given as the functions of time and the number of debugging activities. Section 5 presents several numerical examples of software performance analysis. Finally, Section 6 summarizes the results obtained in this paper.

2. Software Availability Model

2.1. *Model Description*

The following assumptions are made for operational software availability modeling:

A1. The software system breaks down and starts to be restored as soon as a software failure occurs, and the system cannot operate until the restoration action completes.

A2. When a software failure occurs, the restoration action with the debugging activity is performed with probability p $(0 < p < 1)$, on the other hand, without the debugging activity is performed with probability $q(= 1 - p)$.

A3. The debugging activity is perfect with the perfect debugging rate a $(0 < a < 1)$, on the other hand, imperfect with the probability $b(= 1-a)$. If the debugging activity is perfect, one fault is corrected and removed from the system.

A4. When n faults have been corrected, the time to the next software failure-occurrence, U_n, and the restoration time with the debugging activity, L_n^1, follow the exponential distributions with means $1/\lambda_n$ and $1/\mu_n$, respectively. λ_n and μ_n are non-increasing functions of n.

A5. The restoration time without the debugging activity, L_n^2, follows the exponential distribution with mean $1/\eta$.

Let $\{X(t), t \geq 0\}$ be the stochastic process representing the state of the software system at the time point t and its state space is defined as follows:

$\boldsymbol{W} = \{W_n : n = 0, 1, 2, \ldots\}$: the system is operating and available,
$\boldsymbol{R^1} = \{R_n^1 : n = 0, 1, 2, \ldots\}$: the system is inoperable and restored with the debugging activity,

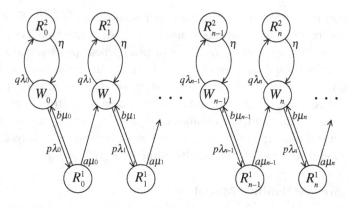

Figure 1. A sample state transition diagram of $X(t)$.

$\boldsymbol{R^2} = \{R_n^2 : n = 0, 1, 2, \ldots\}$: the system is inoperable and restored without the debugging activity,

where n denotes the cumulative number of corrected faults. Figure 1 illustrates the state transition diagram of $X(t)$. Let $Q_{A,B}(t)$ ($A, B \in \{\boldsymbol{W}, \boldsymbol{R^1}, \boldsymbol{R^2}\}$) denote the one-step transition probability that, after making a transition into state A, the process $X(t)$ next makes a transition into state B in an amount of time less than or equal to t. From Fig. 1, we have the following expressions of $Q_{A,B}(t)$'s:

$$Q_{W_n, R_n^1}(t) = p(1 - e^{-\lambda_n t}), \tag{1}$$

$$Q_{W_n, R_n^2}(t) = q(1 - e^{-\lambda_n t}), \tag{2}$$

$$Q_{R_n^1, W_{n+1}}(t) = a(1 - e^{-\mu_n t}), \tag{3}$$

$$Q_{R_n^1, W_n}(t) = b(1 - e^{-\mu_n t}), \tag{4}$$

$$Q_{R_n^2, W_n}(t) = 1 - e^{-\eta t}. \tag{5}$$

2.2. Traditional Software Availability Measures

2.2.1. Distribution of transition time between state $\boldsymbol{W}$

Let $S_{i,n}$ ($i, n = 0, 1, 2, \ldots$; $i \leq n$; $S_{n,n} \equiv 0$) be the random variable representing the transition time of $X(t)$ from state W_i to state W_n, and $G_{i,n}(t)$ be a distribution function of $S_{i,n}$, respectively. Then, we obtain the following renewal equation of $G_{i,n}(t)$:

$$\begin{aligned} G_{i,n}(t) = &Q_{W_i, R_i^1} * Q_{R_i^1, W_{i+1}} * G_{i+1,n}(t) + Q_{W_i, R_i^1} * Q_{R_i^1, W_i} * G_{i,n}(t) \\ &+ Q_{W_i, R_i^2} * Q_{R_i^2, W_i} * G_{i,n}(t) \quad (i = 0, 1, 2, \ldots, n-1), \end{aligned} \tag{6}$$

where $*$ denotes a Stieltjes convolution and $G_{n,n}(t) = 1(t)$ (unit function) ($n = 0, 1, 2, \ldots$). We can solve Eq. (6) with respect to $G_{i,n}(t)$ by applying the Laplace-Stieltjes transform.[15] The solution of Eq. (6) is obtained as

$$G_{i,n}(t) \equiv \Pr\{S_{i,n} \leq t\}$$
$$= 1 - \sum_{m=i}^{n-1} \left[A_{i,n}^1(m)e^{-d_m^1 t} + A_{i,n}^2(m)e^{-d_m^2 t} + A_{i,n}^3(m)e^{-d_m^3 t} \right]$$

$$\left(\begin{array}{l} A_{i,n}^1(m) = \dfrac{\prod_{j=i}^{n-1} pa\lambda_j\mu_j(\eta - d_m^1)}{d_m^1 \prod_{\substack{j=i \\ j \neq m}}^{n-1}(d_j^1 - d_m^1) \prod_{j=i}^{n-1}(d_j^2 - d_m^1)(d_j^3 - d_m^1)} \\[2em] A_{i,n}^2(m) = \dfrac{\prod_{j=i}^{n-1} pa\lambda_j\mu_j(\eta - d_m^2)}{d_m^2 \prod_{\substack{j=i \\ j \neq m}}^{n-1}(d_j^2 - d_m^2) \prod_{j=i}^{n-1}(d_j^3 - d_m^2)(d_j^1 - d_m^2)} \\[2em] A_{i,n}^3(m) = \dfrac{\prod_{j=i}^{n-1} pa\lambda_j\mu_j(\eta - d_m^3)}{d_m^3 \prod_{\substack{j=i \\ j \neq m}}^{n-1}(d_j^3 - d_m^3) \prod_{j=i}^{n-1}(d_j^1 - d_m^3)(d_j^2 - d_m^3)} \\[1em] (m = i, i+1, i+2, \ldots, n-1) \end{array}\right\}, \quad (7)$$

where $-d_m^1$, $-d_m^2$, and $-d_m^3$ are the distinct roots of the following third order equation of s:

$$s^3 + (\lambda_m + \mu_m + \eta)s^2 + [(1-pb)\lambda_m\mu_m + \mu_m\eta + p\eta\lambda_m]s + pa\lambda_m\mu_m\eta = 0, \quad (8)$$

where the proof that Eq. (8) has distinct three negative roots is shown in Appendix A.

Furthermore, the expectation and the variance of $S_{i,n}$ are given by

$$E[S_{i,n}] = \sum_{m=i}^{n-1} \left(\frac{1}{d_m^1} + \frac{1}{d_m^2} + \frac{1}{d_m^3} - \frac{1}{\eta} \right), \quad (9)$$

$$\text{Var}[S_{i,n}] = \sum_{m=i}^{n-1} \left(\frac{1}{(d_m^1)^2} + \frac{1}{(d_m^2)^2} + \frac{1}{(d_m^3)^2} - \frac{1}{\eta^2} \right), \quad (10)$$

respectively.

2.2.2. *Operational state occupancy probability and software availability*

Let $P_{A,B}(t) \equiv \Pr\{X(t) = B | X(0) = A\}$ $(A, B \in \{W, R^1, R^2\})$ be the state occupancy probability that the system is in state B at the time point t on the condition that the system was in state A at time point zero. Then, we obtain the following renewal equation of $P_{W_i, W_n}(t)$:

$$P_{W_i, W_n}(t) = G_{i,n} * P_{W_n, W_n}(t), \tag{11}$$

$$P_{W_n, W_n}(t) = e^{-\lambda_n t} + Q_{W_n, R_n^1} * Q_{R_n^1, W_n} * P_{W_n, W_n}(t) \\ + Q_{W_n, R_n^2} * Q_{R_n^2, W_n} * P_{W_n, W_n}(t). \tag{12}$$

Solving Eqs. (11) and (12), we obtain the operational state occupancy probability as

$$P_{W_i, W_n}(t) \equiv \Pr\{X(t) = W_n | X(0) = W_i\} \\ = \frac{g_{i,n+1}(t)}{pa\lambda_n} + \frac{g'_{i,n+1}(t)}{pa\lambda_n \mu_n}, \tag{13}$$

where $g_{i,n}(t) \equiv dG_{i,n}(t)/dt$ is the density function of $S_{i,n}$ and $g'_{i,n}(t) \equiv d^2 G_{i,n}(t)/dt^2$.

The instantaneous software availability and the average software availability are given by

$$A(t; l) = \sum_{i=0}^{l} \binom{l}{i} a^i b^{l-i} \sum_{n=i}^{\infty} \left[\frac{g_{i,n+1}(t)}{pa\lambda_n} + \frac{g'_{i,n+1}(t)}{pa\lambda_n \mu_n} \right], \tag{14}$$

$$A_{av}(t; l) = \frac{1}{t} \sum_{i=0}^{l} \binom{l}{i} a^i b^{l-i} \sum_{n=i}^{\infty} \left[\frac{G_{i,n+1}(t)}{pa\lambda_n} + \frac{g_{i,n+1}(t)}{pa\lambda_n \mu_n} \right], \tag{15}$$

respectively. Equations (14) and (15) represent the probability that the software system is operable and available at the time point t and the expected proportion of system's operating time to the time interval $(0, t]$, given that the l-th debugging activity ($l = 0, 1, 2, \ldots$) was complete at time point $t = 0$, respectively.

3. Model Analysis

We make the following assumptions for system's task processing.

> B1. The process $\{N(t), t \geq 0\}$ representing the number of tasks arriving at the system up to the time t follows the homogeneous Poisson process with the arrival rate θ.

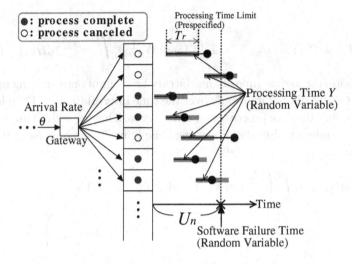

Figure 2. Configuration of task processing.

B2. The processing time of a task, Y, follows a general distribution whose distribution function is denoted as $H(t)$. Each of the processing times is independent.

B3. When the system causes a software failure in task processing or the processing times of tasks exceed the prespecified processing time limit, T_r, the corresponding tasks are canceled.

B4. The number of tasks the system can process simultaneously is sufficiently large.

Figure 2 illustrates the configuration of system's task processing.

Let $\{Z_i(t|T_r), t \geq 0\}$ be the stochastic process representing the cumulative number of tasks whose processes can be complete within the processing time limit T_r out of the tasks arriving up to the time t, given that i faults were corrected and removed from the system at time point $t = 0$. By conditioning with $\{N(t) = k\}$, we obtain the following form of the probability mass function of $Z_i(t|T_r)$:

$$\Pr\{Z_i(t|T_r) = j\} = \sum_{k=0}^{\infty} \Pr\{Z_i(t|T_r) = j | N(t) = k\} e^{-\theta t} \frac{(\theta t)^k}{k!}. \quad (16)$$

From Fig. 2, given that $\{X(t) = W_n\}$, the probability that the process of an arbitrary task is complete within the processing time limit T_r is given by

$$\beta_n(T_r) \equiv \Pr\{Y < U_n,\ Y < T_r | X(t) = W_n\} = \int_0^{T_r} e^{-\lambda_n y} \mathrm{d}H(y). \quad (17)$$

Furthermore, the arrival time of an arbitrary task out of ones arriving up to the time t is distributed uniformly over the time interval $(0,\ t]$.[15] Therefore, the probability that the process of an arbitrary task having arrived up to the time t is complete within the processing time limit T_r, denoted as $\gamma_i(t|T_r)$, is obtained as

$$\gamma_i(t|T_r) = \int_0^t \left(\sum_{n=i}^{\infty} \Pr\{X(x) = W_n | X(0) = W_i\} \times \right.$$
$$\left. \Pr\{Y < U_n,\ Y < T_r | X(x) = W_n\} \right) \frac{\mathrm{d}x}{t}$$
$$= \frac{1}{t} \sum_{n=i}^{\infty} \left[\frac{G_{i,n+1}(t)}{p a \lambda_n} + \frac{g_{i,n+1}(t)}{p a \lambda_n \mu_n} \right] \beta_n(T_r). \quad (18)$$

Then from assumption B2,

$$\Pr\{Z_i(t|T_r) = j | N(t) = k\} =$$
$$\begin{cases} \binom{k}{j} [\gamma_i(t|T_r)]^j [1 - \gamma_i(t|T_r)]^{k-j} & (j = 0,\ 1,\ 2,\ \ldots,\ k) \\ 0 & (j > k) \end{cases}. \quad (19)$$

That is, given that $\{N(t) = k\}$, the number of tasks whose processes can be complete within the processing time limit T_r follows the binomial process with mean $k\gamma_i(t|T_r)$. Accordingly, from Eq. (16) the distribution of $Z_i(t|T_r)$ is given by

$$\Pr\{Z_i(t|T_r) = j\} = \sum_{k=j}^{\infty} \binom{k}{j} [\gamma_i(t)]^j [1 - \gamma_i(t)]^{k-j} e^{-\theta t} \frac{(\theta t)^k}{k!}$$
$$= e^{-\theta t \gamma_i(t|T_r)} \frac{[\theta t \gamma_i(t|T_r)]^j}{j!}. \quad (20)$$

Equation (20) means that $\{Z_i(t|T_r),\ t \geq 0\}$ follows the nonhomogeneous Poisson process with the mean value function $\theta t \gamma_i(t|T_r)$.

Paying attention to the number of incompletable tasks, we can perform the similar analysis. That is, letting $\{V_i(t|T_r),\ t \geq 0\}$ be the stochastic process representing the cumulative number of tasks whose processes cannot be complete within the processing time limit T_r out of the tasks arriving up to the time t, given that i faults were corrected at time point $t = 0$, we obtain the distribution of $V_i(t|T_r)$ as

$$\left. \begin{array}{c} \Pr\{V_i(t|T_r) = j\} = e^{-\theta t \delta_i(t|T_r)} \dfrac{[\theta t \delta_i(t|T_r)]^j}{j!} \\ \delta_i(t|T_r) = 1 - \gamma_i(t|T_r) \end{array} \right\}. \qquad (21)$$

Equation (21) means that $\{V_i(t|T_r),\ t \geq 0\}$ follows the nonhomogeneous Poisson process with the mean value function $\theta t \delta_i(t|T_r)$.

4. Derivation of Software Performance Measures

Based on the above analysis, we can obtain several measures for software performance evaluation considering the real-time property.

The expected numbers of tasks completable and incompletable out of the tasks arriving up to the time t are given by

$$\Lambda_i(t|T_r) \equiv \mathrm{E}[Z_i(t|T_r)] = \theta \sum_{n=i}^{\infty} \left[\frac{G_{i,n+1}(t)}{pa\lambda_n} + \frac{g_{i,n+1}(t)}{pa\lambda_n \mu_n} \right] \beta_n(T_r), \qquad (22)$$

$$\Omega_i(t|T_r) \equiv \mathrm{E}[V_i(t|T_r)] = \theta \left\{ t - \sum_{n=i}^{\infty} \left[\frac{G_{i,n+1}(t)}{pa\lambda_n} + \frac{g_{i,n+1}(t)}{pa\lambda_n \mu_n} \right] \beta_n(T_r) \right\}, \quad (23)$$

respectively.

Furthermore, the instantaneous task completion and incompletion ratios are obtained as

$$\mu_i(t|T_r) \equiv \frac{d\Lambda_i(t|T_r)}{dt} \Big/ \theta = \sum_{n=i}^{\infty} \left[\frac{g_{i,n+1}(t)}{pa\lambda_n} + \frac{g'_{i,n+1}(t)}{pa\lambda_n \mu_n} \right] \beta_n(T_r), \qquad (24)$$

$$\nu_i(t|T_r) \equiv \frac{d\Omega_i(t|T_r)}{dt} \Big/ \theta = 1 - \sum_{n=i}^{\infty} \left[\frac{g_{i,n+1}(t)}{pa\lambda_n} + \frac{g'_{i,n+1}(t)}{pa\lambda_n \mu_n} \right] \beta_n(T_r), \qquad (25)$$

respectively. These equations represent the ratios of the number of tasks whose processes are complete and incomplete within the processing time limit T_r to one arriving at the system per unit time at the time point t, respectively.

As to $\gamma_i(t|T_r)$ in Eq. (18) and $\delta_i(t|T_r)$ in Eq. (21), we can give the following interpretations:

$$\gamma_i(t|T_r) = \frac{\mathrm{E}[Z_i(t|T_r)]}{\mathrm{E}[N(t)]}, \qquad (26)$$

$$\delta_i(t|T_r) = \frac{\mathrm{E}[V_i(t|T_r)]}{\mathrm{E}[N(t)]}, \qquad (27)$$

That is, $\gamma_i(t|T_r)$ and $\delta_i(t|T_r)$ are the cumulative task completion and incompletion ratios up to the time t, respectively.

We should note that it is too difficult to use Eqs. (22)–(27) directly as the software performance measures. The reason is that the cumulative number of faults corrected at the time origin, i.e., integer i cannot be observed immediately since this model assumes the imperfect debugging environment. However, we can easily observe the number of debugging activities and the cumulative number of faults corrected after the completion of the l-th debugging, C_l, is distributed with the probability mass function $\Pr\{C_l = i\} = \binom{l}{i} a^i b^{l-i}$. Similar to the traditional software availability measures in Sec. 2, we can convert Eqs. (22)–(27) into the functions of the number of debuggings, l, i.e., we can obtain

$$\Lambda(t;l|T_r) = \theta \sum_{i=0}^{l} \binom{l}{i} a^i b^{l-i} \sum_{n=i}^{\infty} \left[\frac{G_{i,n+1}(t)}{pa\lambda_n} + \frac{g_{i,n+1}(t)}{pa\lambda_n \mu_n} \right] \beta_n(T_r)$$
$$(l = 0, 1, 2, \ldots), \tag{28}$$

$$\Omega(t;l|T_r) = \theta \left\{ t - \sum_{i=0}^{l} \binom{l}{i} a^i b^{l-i} \sum_{n=i}^{\infty} \left[\frac{G_{i,n+1}(t)}{pa\lambda_n} + \frac{g_{i,n+1}(t)}{pa\lambda_n \mu_n} \right] \beta_n(T_r) \right\}$$
$$(l = 0, 1, 2, \ldots), \tag{29}$$

$$\mu(t;l|T_r) = \sum_{i=0}^{l} \binom{l}{i} a^i b^{l-i} \sum_{n=i}^{\infty} \left[\frac{g_{i,n+1}(t)}{pa\lambda_n} + \frac{g'_{i,n+1}(t)}{pa\lambda_n \mu_n} \right] \beta_n(T_r)$$
$$(l = 0, 1, 2, \ldots), \tag{30}$$

$$\nu(t;l|T_r) = 1 - \sum_{i=0}^{l} \binom{l}{i} a^i b^{l-i} \sum_{n=i}^{\infty} \left[\frac{g_{i,n+1}(t)}{pa\lambda_n} + \frac{g'_{i,n+1}(t)}{pa\lambda_n \mu_n} \right] \beta_n(T_r)$$
$$(l = 0, 1, 2, \ldots), \tag{31}$$

$$\gamma(t;l|T_r) = \frac{1}{t} \sum_{i=0}^{l} \binom{l}{i} a^i b^{l-i} \sum_{n=i}^{\infty} \left[\frac{G_{i,n+1}(t)}{pa\lambda_n} + \frac{g_{i,n+1}(t)}{pa\lambda_n \mu_n} \right] \beta_n(T_r)$$
$$(l = 0, 1, 2, \ldots), \tag{32}$$

$$\delta(t;l|T_r) = 1 - \frac{1}{t} \sum_{i=0}^{l} \binom{l}{i} a^i b^{l-i} \sum_{n=i}^{\infty} \left[\frac{G_{i,n+1}(t)}{pa\lambda_n} + \frac{g_{i,n+1}(t)}{pa\lambda_n \mu_n} \right] \beta_n(T_r)$$
$$(l = 0, 1, 2, \ldots), \tag{33}$$

respectively. Equations (28)–(33) represent the expected numbers of completable and incompletable tasks, the instantaneous task completion and

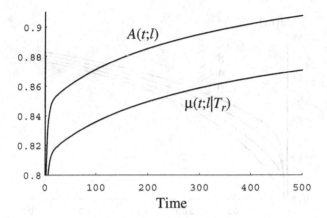

Figure 3. $\mu(t;l|T_r)$ and $A(t;l)$ ($l = 5$; $T_r = 5.0 \times 10^{-3}$, $\nu = 2.0$, $\alpha = 1.0 \times 10^3$, $p = 0.8$, $\eta = 2.0$).

incompletion ratios, and the cumulative task completion and incompletion ratios, given that the l-th debugging was complete at time point $t = 0$, respectively. Equations (30)–(33) have no bearing on the task arrival rate, θ.

5. Numerical Examples

We present several numerical examples on software performance analysis based on the above measures. We apply $\lambda_n \equiv Dc^n$ ($D > 0$, $0 < c < 1$) and $\mu_n \equiv Er^n$ ($E > 0$, $0 < r \leq 1$) to the hazard and the restoration rates, respectively,[16] and cite the estimates of D, c, E, and r from Ref. 17, i.e., we use the following values:

$$\widehat{D} = 0.246, \ \widehat{c} = 0.940, \ \widehat{E} = 1.114, \ \widehat{r} = 0.960,$$

where we set $a = 0.8$.

For the distribution of the processing time, Y, we apply the gamma distribution whose density is given by

$$\frac{dH(t)}{dt} = \frac{\alpha^\nu t^{\nu-1} e^{-\alpha t}}{\int_0^\infty x^{\nu-1} e^{-x} dx} \quad (t \geq 0; \ \nu > 0, \ \alpha > 0), \tag{34}$$

where ν and α are the shape and the scale parameters, respectively. Then the mean and the variance of the processing time are given by $\mathrm{E}[Y] = \nu/\alpha$ and $\mathrm{Var}[Y] = \nu/\alpha^2$, respectively.

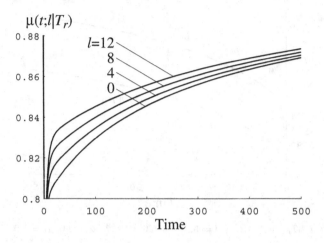

Figure 4. $\mu(t;l|T_r)$ for various numbers of debuggings, l ($T_r = 5.0 \times 10^{-3}$, $\nu = 2.0$, $\alpha = 1.0 \times 10^3$, $p = 0.8$, $\eta = 2.0$).

Figure 3 shows the time-dependent behaviors of the instantaneous task completion ratio, $\mu(t;l|T_r)$, in Eq. (30) and the instantaneous software availability, $A(t;l)$, in Eq. (14). This figure tells us that the new measure considering the real-time property ($\mu(t;l|T_r)$) gives more pessimistic evaluation than the traditional one ($A(t;l)$).

Figures 4 and 5 show $\mu(t;l|T_r)$ and the cumulative task completion ratio, $\gamma(t;l|T_r)$, in Eq. (32) for various numbers of debuggings, l, respectively. As shown in these figures, we can see that the system performance also improves as the debugging is progressing and that the proposed quantities enable us to understand the relationship between the software performance evaluation and the number of debugging activities.

Figure 6 show the dependence of $\gamma(t;l|T_r)$ on the value of p, representing the probability that the debugging activity is performed when the system is down. This figure indicates that software performance is evaluated lower in the early stage of the operation phase but more improves with the lapse of time as the value of p increases. The larger value of p gives the following two impacts: (i) software reliability growth occurs earlier, on the other hand, (ii) the unavailable (restoration) time tends to be longer since the mean restoration time with debugging, $\mathrm{E}[L_n^1] = 1/\mu_n$, is assumed to be the increasing function of n. As to the larger p, impact (ii) appears in the early stage of the operation phase and then impact (i) becomes larger gradually with the lapse of time.

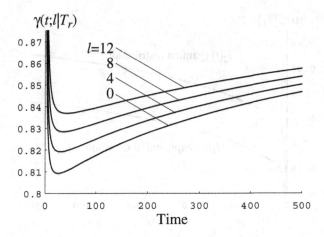

Figure 5. $\gamma(t;l|T_r)$ for various numbers of debuggings, l ($T_r = 5.0 \times 10^{-3}$, $\nu = 2.0$, $\alpha = 1.0 \times 10^3$, $p = 0.8$, $\eta = 2.0$).

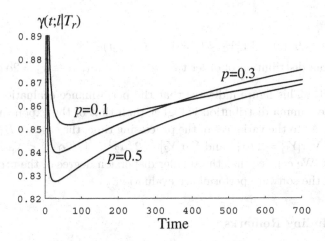

Figure 6. Dependence of $\gamma(t;l|T_r)$ on p ($l = 0$; $T_r = 5.0 \times 10^{-3}$, $\nu = 2.0$, $\alpha = 1.0 \times 10^3$, $\eta = 2.0$).

Figure 7 shows the dependence of $\mu(t,l|Tr)$ on the distribution of the processing time of a task, $H(t)$. In Fig. 7, we set the parameters ν and α as equalize the means of the processing time for $H_1(t)$ and $H_2(t)$, i.e.,

$$H(t) \equiv H_1(t) = \Pr\{Y_1 \le t\} = 1 - e^{-\alpha_1 t}$$

(exponential distribution: $\nu \equiv \nu_1 = 1.0$, $\alpha \equiv \alpha_1 = 500.0$), (35)

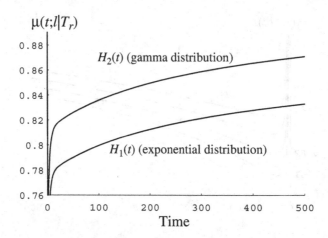

Figure 7. Dependence of $\mu(t; l|T_r)$ on distribution of Y, $H(t)$ ($l = 5$; $T_r = 5.0 \times 10^{-3}$, $p = 0.8$, $\eta = 2.0$).

$$H(t) \equiv H_2(t) = \Pr\{Y_2 \leq t\} = 1 - (1 + \alpha_2 t)e^{-\alpha_2 t}$$

(gamma distribution of order two: $\nu \equiv \nu_2 = 2\nu_1$, $\alpha \equiv \alpha_2 = 2\alpha_1$), (36)

respectively. This figure indicates that the performance evaluation in the case of the gamma distribution is higher than that of the exponential distribution. As to the variances of the processing time, the cases of $H_1(t)$ and $H_2(t)$ are $\mathrm{Var}[Y_1] = 1/\alpha_1^2$ and $\mathrm{Var}[Y_2] = 2/\alpha_2^2 = 1/(2\alpha_1^2) < \mathrm{Var}[Y_1]$, respectively. We can see that the smaller dispersion-degree of the processing time rises the software performance evaluation.

6. Concluding Remarks

In this paper, we have developed the operation-oriented performance-evaluation model for multi-task software system considering the software reliability growth process and the upward tendency of difficulty in debugging. In particular, we have considered two different kinds of restoration actions performed during the operation phase, i.e., the restoration action with or without debugging. The dynamic behavior of the software system itself alternating between up and down states has been described by a Markov process. The distribution of the number of tasks whose processes can be complete within a prespecified processing time limit has been formulated with the infinite-server queueing model. Several useful stochastic quantities for measuring the real-time property of the software system have been derived. These have been given as the functions of time and the number of debuggings. Numerical illustrations for software performance measurement and assessment have been also presented to show that these measures are very useful for operational real-time property evaluation of software systems.

Acknowledgments

This work was supported in part by Grants-in-Aid for Scientific Research (C) of the Ministry of Education, Culture, Sports, Science and Technology of Japan under Grant No. 18510124.

Appendix A. Proof that Eq. (8) Has Distinct Three Negative Roots

Proof. It is natural that $\lambda_m < \mu_m < \eta$ is assumed since the relationship among the expectations of U_m, L_m^1, and L_m^2 is $1/\eta < 1/\mu_m < 1/\lambda_m$ in the general case.

As a preparation, we consider the following function of s:

$$f_0(s) = (s + \lambda_m)(s + \mu_m)(s + \eta)$$
$$= s^3 + (\lambda_m + \mu_m + \eta)s^2 + (\lambda_m\mu_m + \mu_m\eta + \eta\lambda_m)s + \lambda_m\mu_m\eta. \quad (A.1)$$

Obviously, equation $f_0(s) = 0$ has the distinct three roots, $s = -\eta$, $-\mu_m$, $-\lambda_m$, and the behavior of $f_0(s)$ is summarized in Table A.

Table A. Behavior of $f_0(s)$.

s	...	$-\eta$	...	$-\mu_m$	...	$-\lambda_m$	...	0
Sign of $f_0(s)$	−	0	+	0	−	0	+	

Now we consider the function $f(s)$ denoted as

$$f(s) = s^3 + (\lambda_m + \mu_m + \eta)s^2 + [(1 - pb)\lambda_m\mu_m + \mu_m\eta + p\eta\lambda_m]s$$
$$+ pa\lambda_m\mu_m\eta, \quad (A.2)$$

and the difference of $f_0(s)$ and $f(s)$

$$f_0(s) - f(s) = [pb\mu_m + (1-p)\eta]\lambda_m s + (1 - pa)\lambda_m\mu_m\eta. \quad (A.3)$$

From Eq. (A.3), $f_0(s) > f(s)$ in $s \in (-s_0, \infty)$ and $f_0(s) < f(s)$ in $s \in (-\infty, -s_0)$, where

$$-s_0 = \frac{-(1-pa)\mu_m \eta}{pb\mu_m + (1-p)\eta}. \qquad (A.4)$$

We note that $-s_0$ exists in $-\eta < -s_0 < -\mu_m$ because

$$-\mu_m - (-s_0) = \frac{pb\mu_m(\eta - \mu_m)}{pb\mu_m + (1-p)\eta} > 0, \qquad (A.5)$$

$$-s_0 - (-\eta) = \frac{(1-p)\eta(\eta - \mu_m)}{pb\mu_m + (1-p)\eta} > 0. \qquad (A.6)$$

From the above mentioned and Table A, we can see $f(0) > 0$, $f(-\lambda_m) < 0$, $f(-\mu_m) < 0$, and $f_0(-s_0) = f(-s_0) > 0$. Accordingly, equation $f(s) = 0$ has at least one negative root in $(-\lambda_m, 0)$ and $(-s_0, -\mu_m)$, respectively. Furthermore, since $f(-\eta) > 0$ and $\lim_{s \to -\infty} f(s) = -\infty$, $f(s) = 0$ has at least one negative root in $(-\infty, -\eta)$. Considering that any third order equation has at most three roots, it is proved that equation $f(s) = 0$ has only one root in the respective ranges of $(-\infty, -\eta)$, $(-s_0, -\mu_m)$, and $(-\lambda_m, 0)$, i.e., Eq. (8) has distinct three negative roots. □

References

1. M. Tortorella, Service reliability theory and engineering, I: Foundations, *Quality Technology and Quantitative Management*, **2** (1), 1–16, (2005).
2. M. Tortorella, Service reliability theory and engineering, II: Models and examples, *Quality Technology and Quantitative Management*, **2** (1), 17–37, (2005).
3. H. Asama, Service engineering and system integration, *Journal of the Society of Instrument and Control Engineering*, **44** (4), 278–283, (2005) (in Japanese).
4. H. Mizuta, Emergence of service science: Services sciences, management and engineering (SSME), *IPSJ Magazine*, **47** (5), 457–472, (2006) (in Japanese).
5. http://www.saforum.org.
6. M.D. Beaudry, Performance-related reliability measures for computing systems, *IEEE Transactions of Computers*, **C-27** (6), 540–547, (1978).
7. J.F. Meyer, On evaluating the performability of degradable computing systems, *IEEE Transactions of Computers*, **C-29** (8), 720–731, (1980).
8. H. Pham, *System Software Reliability*, Springer-Verlag, London, (2006).
9. S. Yamada, Software reliability models, In *Stochastic Models in Reliability and Maintenance*, (Edited by S. Osaki), pp. 253–280, Springer-Verlag, Berlin, (2002).

10. M. Kimura and S. Yamada, Performance evaluation modeling for redundant real-time software systems, *Transactions of IEICE*, **J78-D-I** (8), 708–715, (1995) (in Japanese).
11. M. Kimura, M. Yamamoto, and S. Yamada, Performance evaluation modeling for fault-tolerant software systems with processing time limit, *Journal of Reliability Engineering Association of Japan*, **20** (7), 422–432, (1998) (in Japanese).
12. J.K. Muppala, S.P. Woolet, and K.S. Trivedi, Real-time-systems performance in the presence of failures, *Computer*, **24** (5), 37–47, (1991).
13. H. Ihara, A review of real time systems, *Journal of Information Processing Society of Japan*, **35** (1), 12–17, (1994) (in Japanese).
14. K. Tokuno and S. Yamada, Operational software availability measurement with two kinds of restoration actions, *Journal of Quality in Maintenance Engineering*, **4** (4), 273–283, (1998).
15. S. Osaki, *Applied Stochastic System Modeling*, Springer-Verlag, Berlin, (1992).
16. P.B. Moranda, Event-altered rate models for general reliability analysis, *IEEE Transactions on Reliability*, **R-28** (5), 376–381, (1979).
17. K. Tokuno and S. Yamada, Stochastic performance evaluation for multi-task processing system with software availability model, *Journal of Quality in Maintenance Engineering*, **12** (4), 412–424, (2006).

SOFTWARE RELIABILITY ASSESSMENT WITH 2-TYPES IMPERFECT DEBUGGING ACTIVITIES*

SHINJI INOUE AND SHIGERU YAMADA

Department of Social Systems Engineering,
Faculty of Engineering,
Tottori University,
4-101 Minami, Koyama-cho, Tottori, Tottori 680-8552, Japan
ino@sse.tottori-u.ac.jp

A software reliability growth model (SRGM) which is known as one of the useful mathematical tools to assess software reliability quantitatively can be classified into two models: Perfect and imperfect debugging models. In an actual testing phase, we can consider that the debugging activities do not always remove faults perfectly. Therefore, the imperfect debugging model is an ideal one for practical software reliability assessment. Under imperfect debugging environment, we can consider two kinds of imperfect debugging activities, such as the activities inducing new fault-introduction and the imperfect fault-correction activities. In this paper, we discuss software reliability growth modeling considering with such two kinds of imperfect debugging activities, and show numerical examples of our imperfect debugging models by using actual fault count data.

1. Introduction

Software reliability assessment conducted in a testing-phase of a software development process is one of the important project management activities to produce highly reliable software systems. A software reliability growth model (abbreviated as SRGM) [1,2,3,4] has been known as one of the fundamental technologies for assesssing software reliability quantitatively, and is applied for practical use. The SRGM is a mathematical tool based on probability and statistical theories. And, it is said that the SRGMs proposed so far are divided into the following categories [5]: Software failure-

---

*This work was supported in part by the Grant-in-Aid for Young Scientists (B), Grant No. 19710129, and the Grant-in-Aid for Scientific Research (C), Grant No. 18510124, from the Ministry of Education, Sports, Science, and Technology of Japan. This work was also conducted as a part of the 2007 Research and Development Project of the Tottori University Venture Business Laboratory.

occurrence time model, software fault-detection count model, and software availability model. Especially the software fault-detection count model can describe a software reliability growth process by regarding the number of faults detected up to arbitrary testing-time as a random variable. Therefore, a nonhomogeneous Poisson process (abbreviated as NHPP) as one of the counting processes is often applied to software fault-detection count modeling.

Further, software fault-detection count models can be classified into two kinds of model: Perfect and imperfect debugging models. The perfect debugging models can be modeled by assuming a perfect debugging environment in which faults latent in a software system are always detected and removed perfectly by the debugging activities. On the other hand, the imperfect debugging models are developed by assuming an imperfect debugging environment where faults are not always detected and removed perfectly and there is a possibility that new faults are introduced by the debugging activities. We can see that the imperfect debugging environment has a suitable assumption for software reliability growth modeling since actual debugging activities can not always detect and remove faults perfectly in the testing-phase. Therefore, a lot of SRGMs developed under some assumptions on the imperfect debugging environment have been proposed so far. Yamada et al. [6], Zeephongsekul [7], Yamada [8], and Yamada and Sera [9] have been proposed several types of imperfect debugging models based on NHPPs, respectively, by considering that new faults are introduced by imperfect debugging activities, respectively. And Yamada et al. [10] has extended a well-known Goel-Okumoto model to an imperfect debugging model by using a perfect debugging rate. Thus, imperfect debugging models proposed so far treat the following two types of imperfect debugging activities separately: (1) imperfect fault-correction activities which introduce new faults, (2) imperfect fault-correction activities which introduce no new faults.

However, it is possible that these two kinds of imperfect debugging activities often occur simultaneously in the actual testing-phase. Accordingly, we need to consider such two kinds of imperfect debugging activities simultaneously for developing plausible imperfect debugging models. In this paper, we develop imperfect debugging models which incorporate such two kinds of imperfect debugging activities simultaneously based on the NHPPs. Especially, we develop two types of imperfect debugging models by assuming two functions describing the time-dependent behavior of the expected total numbers of faults in a software system, which take the numbers of introduced faults into considerations, respectively. Then, we derive several software reliability assessment measures for each model, such as software reliability functions, hazard rates, and mean time between software failures based on the stochastic properties of our imperfect debugging models. Further, we discuss parameter estimation, and show numerical examples of our model by using actual fault count data.

2. Imperfect Debugging Modeling

2.1. NHPP model

We develop imperfect debugging models based on NHPP. An NHPP model is one of the SRGMs, and follows the following counting process, $\{N(t), t \geq 0\}$, representing the number of faults detected during a constant time-interval $(0, t]$:

$$\Pr\{N(t) = n\} = \frac{\{H(t)\}^n}{n!} \exp\{-H(t)\} \quad (1)$$

$$(n = 0, 1, 2, \cdots),$$

where $H(t)$ is called a mean value function which represents the expected number of faults detected during constant time-interval $(0, t]$ and $H(t) \equiv E[N(t)]$. The stochastic behavior of the fault-detection phenomenon can be characterized by assuming a suitable mean value function $H(t)$. Almost all of the mean value functions are developed by assuming basically that the number of faults detected at testing-time t is proportional to the current residual fault content [3].

2.2. Modeling with two kinds of imperfect debugging activities

Assuming that the fault-detection phenomenon in the testing-phase follows the NHPP in Eq. (1), we develop a mean value function by considering two kinds of imperfect debugging activities, such as the activities which introduce new faults and those which introduce no new faults, simultaneously. Our imperfect debugging models are developed by the following assumption on the time-dependent behavior of the expected number of faults detected at testing-time t:

$$\frac{dH(t)}{dt} = b(t)[a(t) - pH(t)] \quad (a(t) \geq a,\ 0 < p \leq 1), \quad (2)$$

where p and $a(t)$ represent the perfect debugging rate and the number of faults in the software system at testing-time t by incorporating the effect of the imperfect debugging activities which introduce new faults, respectively. $b(t)$ the fault-detection rate at testing-time t, and is positive. Accordingly, $(1-p)$ denotes the probability of the imperfect debugging which introduces no new faults. Solving Eq. (2) with respect to $H(t)$, we obtain the following equation:

$$\left. \begin{array}{l} H(t) = e^{-p \cdot Z(t)} \left[\int_0^t a(s)b(s)e^{p \cdot Z(s)} ds \right] \\ Z(t) = \int_0^t b(s) ds \end{array} \right\}. \quad (3)$$

We need to give suitable functions to $a(t)$ and $b(t)$ in Eq. (3), respectively, for developing a specific imperfect debugging model. However, as to the function $a(t)$, it is very difficult to observe the time-dependent behavior of the newly introduced from fault count data or software failure-occurrence times data which are typical data collected normally from actual testing-phase. Therefore, as particular cases of $a(t)$, we consider the following two types of functions which describe the time-dependent behavior of the expected total numbers of faults in the software system by taking the numbers of introduced faults into consideration:

$$a_1(t) = \alpha_1 \exp[\beta t] \quad (\beta > 0), \quad (4)$$

$$a_2(t) = \alpha_2(1 + \gamma t) \quad (\gamma > 0), \quad (5)$$

respectively. In Eqs. (4) and (5), $\alpha_i (i = 1, 2)$ represent the expected numbers of initial inherent faults, β and γ new fault-introduction rates for the expected numbers of initial inherent faults, respectively. We can see that Eqs. (4) and (5) imply that the expected numbers of faults in the software system increase exponentially and linearly with constant increasing rates, respectively. Substituting Eqs. (4) and (5) into Eq. (3) and solving Eq.(3), we can obtain the following mean value functions of the NHPPs:

$$H_1(t) = \frac{\alpha_1 b_1}{pb_1 + \beta}(\exp[\beta t] - \exp[-pb_1 t]), \quad (6)$$

$$H_2(t) = \frac{\alpha_2}{p} \left\{ (1 - \frac{\gamma}{pb_2})(1 - \exp[-pb_2 t]) + \gamma t \right\}, \quad (7)$$

respectively, where we set $b(t) \equiv b_i (i = 1, 2)$ (positive constant values). In this paper we call the NHPP models with Eqs. (6) and (7) as Model 1 and Model 2, respectively.

In Eqs. (6) and (7), we can see that the mean value functions, $H_i(t)(i = 1, 2)$, have the following properties:

$$\left.\begin{array}{l} H_1(0) = H_2(0) = 0 \\ H_1(\infty) = H_2(\infty) = \infty \end{array}\right\}. \qquad (8)$$

That is, the numbers of faults detected in infinitely long duration are infinity, respectively, since these functions assume that new faults are introduced when debugging activities are conducted. In existing imperfect debugging models, Littlewood-Verrall model [11], Weibull process model [1], logarithmic Poisson execution time model [1] have the same properties as Eq. (8) especially.

3. Software Reliability Assessment Measures

We derive software reliability assessment measures of our imperfect debugging models, such as software reliability functions and hazard rates, based on the stochastic properties of our models. These measures are useful metrics for quantitative assessment of software reliability.

3.1. *Software reliability function*

The software reliability function is one of the well-known software reliability assessment measures. Given that the testing or the operation has been going up to testing-time t, the software reliability function is defined as the probability that a software failure does not occur in the time-interval $(t, t + x] (t \geq 0, x \geq 0)$. Accordingly, the software reliability function $R(x \mid t)$ can be formulated as

$$R(x \mid t) \equiv \exp[-\{H(t + x) - H(t)\}], \qquad (9)$$

which is derived from the properties of the NHPP. Substituting Eqs. (6) and (7) into Eq. (9), we have the software reliability functions as

$$R_1(x \mid t) = \exp\left[-\frac{\alpha_1 b_1}{pb_1 + \beta}(\exp[\beta(t + x)]\right.$$
$$\left. - \exp[-pb_1(t + x)] - \exp[\beta t]) + \exp[-pb_1 t]\right], \qquad (10)$$

$$R_2(x \mid t) = \exp\left[\frac{\alpha_2}{p}\left\{(1 - \frac{\gamma}{pb_2})(\exp[-pb_2(t + x)] - \exp[-pb_2 t]) - \gamma t\right\}\right], \qquad (11)$$

respectively.

3.2. Mean time between software failures

Let $F(x \mid t)$ be the probability that a software failure occurs in the time-interval $(t, t + x]$. Then, we can see that our imperfect debugging models conserve the following properties: $F(0 \mid t) = 0$ and $F(\infty \mid t) = 1$ because of Eq. (8). That is, the probability distribution function, $F(x \mid t)$, satisfies the properties of the ordinary probability distribution function. Accordingly, we can derive mean time between software failure (MTBF) of our each model. The MTBF is formulated as

$$E(X \mid t) = \int_0^\infty R(x \mid t) dx. \tag{12}$$

By Eq. (12), we can derive MTBFs of each our imperfect debugging model.

3.3. Hazard rate

The hazard rate represents the frequency of software failure-occurrence per unit testing-time, and is formulated as

$$z(x \mid t) \equiv \frac{-\frac{d}{dx} R(x \mid t)}{R(x \mid t)} = h(t + x), \tag{13}$$

where $z(x \mid t) \Delta t$ represents the probability that a software failure occurs during a small time-interval $(t + x, t + x + \Delta t]$ given that a software failure has not been occurring during a time-interval $(t, t + x] (t \geq 0, x \geq 0)$. Substituting Eqs. (6) and (7) into Eq. (13), we obtain the following hazard rate functions:

$$z_1(x \mid t) = \frac{\alpha_1 b_1}{pb_1 + \beta}(\beta \exp[\beta(t + x)] + pb_1 \exp[-pb_1(t + x)]), \tag{14}$$

$$z_2(x \mid t) = \frac{\alpha_2}{p}\{\exp[-pb_2(t + x)](pb_2 - \gamma) + \gamma\}, \tag{15}$$

respectively.

4. Parameter Estimation

We discuss the methods of parameter estimation for our imperfect debugging models. As the first step, we discuss the estimation methods for the parameters related to the imperfect debugging activities, such as p, β, and γ in Eqs. (6) and (7). Generally, estimating these parameters related to the imperfect debugging activities by using fault count data or software failure-occurrence times data is very difficult. Therefore, we need to give the values of these parameters experimentally in this case. In this paper, we calculate the values of the parameters β and γ by using fault-introduction rates [6], $c_i (i = 1, 2)$, which are formulated as

$$c_i = \frac{a_i(T) - \alpha_i}{\alpha_i} \quad (i = 1, 2), \tag{16}$$

where the subscripts, $i = 1, 2$, indicate the types of our imperfect debugging models, Eqs. (6) and (7), respectively. In Eq. (16), T denotes the termination time of the testing. By substituting Eqs. (4) and (5) into Eq. (16), the fault-introduction rates can be rewritten as

$$c_1 = \exp[\beta T] - 1, \tag{17}$$

$$c_2 = \gamma T, \tag{18}$$

respectively. From Eqs. (17) and (18), the parameters β and γ can be calculated as

$$\bar{\beta} = \frac{1}{T} \log(\bar{c}_1 + 1), \tag{19}$$

$$\bar{\gamma} = \frac{1}{T} \bar{c}_2, \tag{20}$$

by giving the values of the fault-introduction rates as $\bar{c}_i (i = 1, 2)$, respectively.

Next we discuss methods of parameter estimation for α_i and b_i ($i = 1, 2$). In this paper we estimate the parameters α_i and b_i ($i = 1, 2$) based on the method of maximum-likelihood by using the set and calculated parameters p, β, and γ. Supposing that K data pairs $(t_k, y_k)(k = 0, 1, 2, \cdots, K)$ have been observed with respect to the cumulative number of faults, y_k, detected during a constant time-interval $(0, t_k](0 < t_1 < t_2 < \cdots < t_K)$, we can derive the likelihood functions, L, as:

$$L = \Pr\{N(t_1) = y_1, N(t_2) = y_2, \cdots, N(t_K) = y_K\}$$

$$= \exp[-H_i(t_K)] \prod_{k=1}^{K} \frac{\{H_i(t_k) - H_i(t_{k-1})\}^{y_k - y_{k-1}}}{(y_k - y_{k-1})!} \quad (i = 1, 2). \tag{21}$$

Then, the logarithmic likelihood functions are derived as

$$\ln L_i = \sum_{k=1}^{K} (y_k - y_{k-1}) \cdot \ln[H_i(t_k) - H_i(t_{k-1})] - H_i(t_K)$$

$$- \sum_{k=1}^{K} \ln[y_k - y_{k-1}] \quad (i = 1, 2), \tag{22}$$

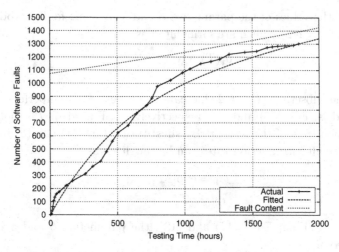

Figure 1. The estimated mean value function, $\widehat{H}_1(t)$. (Model 1 ; $p = 0.95$, $\bar{c}_1 = 0.3$, $\beta = 1.421 \times 10^{-4}$).

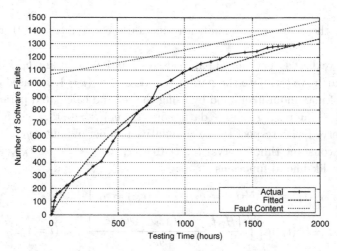

Figure 2. The estimated mean value function, $\widehat{H}_2(t)$. (Model 2 ; $p = 0.95$, $\bar{c}_2 = 0.3$, $\gamma = 1.624 \times 10^{-4}$).

by using the properties of NHPPs [3,12]. Accordingly, the parameter estimates $\widehat{\alpha}_i$ and $\widehat{b}_i (i = 1, 2)$ of parameters α_i and b_i can be obtained by solving the simultaneous logarithmic likelihood equations derived from Eq. (22), with respect to the parameters α_i and b_i ($i = 1, 2$), respectively.

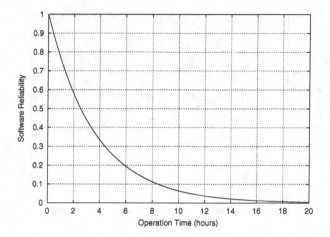

Figure 3. The estimated software reliability function, $\widehat{R}_1(x \mid 1846.92)$.

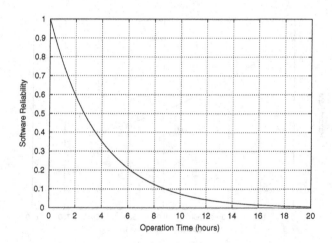

Figure 4. The estimated software reliability function, $\widehat{R}_2(x \mid 1846.92)$.

5. Numerical Examples

We show numerical examples of our imperfect debugging models by using actual fault count data. The actual data set to be used in this paper consists of 35 data pairs $(t_k, y_k)(k = 1, 2, \cdots, 35; t_{35} = 1846.92$ (hours)$, y_{35} = 1301)$ [13]. In this paper, we set $p = 0.95$ and $\bar{c}_i = 0.3$ ($i = 1, 2$). Then, we can calculate $\beta = 1.421 \times 10^{-4}$ and $\gamma = 1.624 \times 10^{-4}$ by using Eqs. (19) and (20), respectively. Using these parameters specified as above, we can

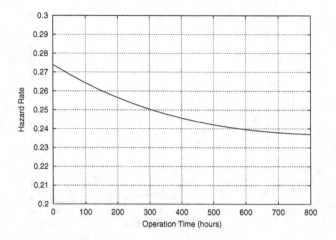

Figure 5.   The estimated hazard rate function, $\widehat{z}_1(x \mid 1846.92)$.

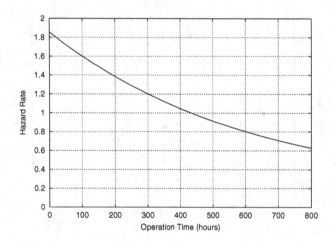

Figure 6.   The estimated hazard rate function, $\widehat{z}_2(x \mid 1846.92)$.

obtain parameter estimates $\widehat{a}_1 = 1.073 \times 10^3$, $\widehat{b}_1 = 1.730 \times 10^{-3}$, $\widehat{a}_2 = 1.066 \times 10^3$, and $\widehat{b}_2 = 1.740 \times 10^{-3}$ by using the method of maximum-likelihood, respectively.

Figures 1 and 2 depict the estimated mean value functions, $\widehat{H}_1(t)$ and $\widehat{H}_2(t)$, respectively. And, Figures 3 and 4 show the estimated software reliabilities $\widehat{R}_1(x \mid 1846.92)$ and $\widehat{R}_2(x \mid 1846.92)$ after the termination time of the testing ($t_{35} = 1846.92$ (hours)), respectively. If we assume that the developed software system is used in the operational phase like in the testing-phase, we can estimate the software reliabilities after 5 hours from the termination time of the testing $\widehat{R}_1(5 \mid 1846.92)$ and $\widehat{R}_2(5 \mid 1846.92)$ to be about 0.255 and 0.271 from Figures 3 and 4, respectively. Furthermore, Figures 5 and 6 show the estimated hazard rates, $\widehat{z}_1(x \mid 1846.92)$ and $\widehat{z}_2(x \mid 1846.92)$, respectively. From Figures 5 and 6, we can estimate the hazard rates after 800 hours from the termination time of the testing $\widehat{z}_1(800 \mid 1846.92)$ and $\widehat{z}_2(800 \mid 1846.92)$ to be about 0.237 and 0.628, respectively.

6. Model Comparison

We compare our two types of imperfect debugging models in terms of mean square errors (abbreviated as MSE). The MSE is calculated by dividing the sum of squared vertical distance between the observed and estimated expected cumulative numbers of faults, y_k and $\widehat{H}(t_k)$, detected during the time-interval $(0, t_k]$, respectively, by the number of observed data pairs. That is, supposing that K data pairs (t_k, y_k) $(k = 1, 2, \cdots, K)$ are observed, we can formulate the MSE as

$$\text{MSE} = \frac{1}{K} \sum_{k=1}^{K} \left[y_k - \widehat{H}(t_k) \right]^2, \qquad (23)$$

where $\widehat{H}(t_k)$ denotes the estimated value of the expected cumulative number of faults by testing-time $t_k (k = 1, 2, \cdots, K)$. The model having the smallest value of the MSE fits best to the observed data set.

In these model comparisons, the parameters are set $p = 0.95, \beta = 1.421 \times 10^{-4}$, and $\gamma = 1.624 \times 10^{-4}$ which are the same as in Section 5, and we use the following fault count data collected in actual software project:

- DS1 [13] : $(t_k, y_k)(k = 1, 2, \cdots, 35$; $t_{35} = 35, y_{35} = 1301)$ where t_k is measured on the basis of hours.
- DS2 [13] : $(t_k, y_k)(k = 1, 2, \cdots, 19$; $t_{19} = 19, y_{19} = 328)$ where t_k is measured on the basis of weeks.
- DS3 [14] : $(t_k, y_k)(k = 1, 2, \cdots, 24$; $t_{24} = 24, y_{24} = 296)$ where t_k is measured on the basis of weeks.
- DS4 [15] : $(t_k, y_k)(k = 1, 2, \cdots, 59$; $t_{59} = 59, y_{59} = 5186)$ where t_k is measured on the basis of weeks.

Table 1. First five normalized natural frequencies. The discovery of a very small kinetic energy.

	DS1	DS2	DS3	DS4
Model 1	4051.39	243.375	690.438	49086.4
Model 2	**3913.79**	**237.192**	**674.719**	**45486.4**

Table 1 shows the results of model comparisons based on the MSE. In Table 1, we can say that Model 2 has better performance than Model 1 in terms of the MSE when we set the parameter as $p = 0.95$, $\beta = 1.421 \times 10^{-4}$, and $\gamma = 1.624 \times 10^{-4}$. That is, it is shown that it is better to assume a linearly increasing function representing the expected number of faults considering with new introduced faults, such as Eq. (5), in these model comparisons.

7. Concluding Remarks

We have proposed two types of imperfect debugging models based on the NHPPs by considering with simultaneous two kind of imperfect debugging activities, such as the activities which introduce new faults and those which introduce no new faults. Then, we have derived software reliability assessment measures based on the stochastic properties of the NHPPs. Finally, we have discussed the method of parameter estimation, and shown numerical examples for our imperfect debugging models by using actual fault count data. In further studies, we plan to develop more plausible imperfect debugging models which can describe the software reliability growth process with the two kinds of imperfect debugging activities by using other suitable stochastic processes, and have to research in a method to estimate the values of the parameter related to the imperfect debugging activities, evaluate the validity and usefulness of our models for practical software reliability assessment.

References

1. J.D. Musa, D. Iannio, and K. Okumoto, *Software Reliability: Measurement, Prediction, Application.* McGraw-Hill, New York (1987).
2. H. Pham, *Software Reliability.* Springer-Verlag, Singapore (2000).
3. S. Yamada, Software reliability models, in *Stochastic Models in Reliability and Maintenance* (S. Osaki, Ed.), Springer-Verlag, Berlin, 253–280 (2002).
4. S. Yamada and S. Osaki, Software reliability growth modeling: Models and applications, *IEEE Trans. Softw. Eng.*, **SE-11**(12), 1431–1437 (1985).

5. S. Yamada, Software reliability models and their applications: A survey, *Intern. Semi. Soft. Rel. Man-Ma. Sys.*, 56–80 (2000).
6. S. Yamada, K. Tokuno, and S. Osaki, Imperfect debugging models with fault introduction rate for software reliability assessment, *Int. J. Syst. Sci.*, **23**(12), 2241–2252 (1991).
7. P. Zeephongsekul, Reliability growth of a software model under imperfect debugging and generation of errors, *Micro. & Reliab.*, **36**(10), 1475–1482 (1996).
8. S. Yamada, Software reliability growth models incorporating imperfect debugging with introduced faults, *Elect. and Comm. in Japan (Part 3)*, **84**(4), 33–41 (1998).
9. S. Yamada and K. Sera, Imperfect debugging models with two kinds of software hazard rate and their Bayesian formulation, *Elect. and Comm. in Japan (Part 3)*, **84**(3), 12–20 (2001).
10. S. Yamada, T. Yamane, and S. Osaki, Software reliability growth models with error debugging rate, (in Japanese) *Trans. IPS Japan*, **27**(1), 64–71 (1986).
11. B. Littlewood and J.L. Verrall, A Bayesian reliability growth model for computer software," *J. Royal of Stat. Socie. Ser. C*, **22**(3), 332–346 (1973).
12. S. Osaki, *Applied Stochastic System Modeling*. Springer-Verlag, Berlin, Heidelberg (1992).
13. W.D. Brooks and R.W. Motley, Analysis of Discrete Software Reliability Models," Techn. Rep. RADC-TR-80–84, Rome Air Development Center, New York (1980).
14. T. Fujiwara and S. Yamada, C0 coverage-measure and testing-domain metrics based on a software reliability growth model," *Int. J. Reliab., Qual. Saf. Eng.*, **9**(4), 329–340 (2002).
15. D. Satoh, A discrete Gompertz equation and a software reliability growth model," *IEICE Trans. Inf. and Syst.*, **E83-D**(7), 1508–1513 (2000).

FLEXIBLE STOCHASTIC DIFFERENTIAL EQUATION MODELING FOR OPEN-SOURCE-SOFTWARE RELIABILITY ASSESSMENT

YOSHINOBU TAMURA[†] AND SHIGERU YAMADA[††]

[†] *Department of Computer Science*
Faculty of Applied Information Science
Hiroshima Institute of Technology
Miyake 2-1-1, Saeki-ku, Hiroshima-shi, 731-5193 Japan
tam@cc.it-hiroshima.ac.jp

[††] *Department of Social Systems Engineering*
Faculty of Engineering
Tottori University
Minami 4-101, Koyama, Tottori-shi, 680-8552 Japan
yamada@sse.tottori-u.ac.jp

All over the world people can gain information at the same time by growing rate of Internet access around the world in recent years. In accordance with such a penetration of the Internet, it is increasing public awareness of the importance of online real-time and interactive functions. Therefore, software development environment has been changing into new development paradigms such as concurrent distributed development environment and so-called open source project by using network computing technologies. Especially, such OSS (Open Source Software) systems which serve as key components of critical infrastructures in the society are still ever-expanding now. In this paper, we propose a software reliability growth model based on stochastic differential equations in order to consider the active state of open source project. Especially, we assume that the software failure intensity depends on the time, and the software fault-reporting phenomena on the bug tracking system keep an irregular state. Moreover, we assume that the software failure intensity depends on the fault importance levels. Also, we analyze actual software fault-count data to show numerical examples of software reliability assessment. We show that our proposed model can assist improvement of quality for OSS systems.

1. Introduction

Network technologies have made rapid progress with the dissemination of computer systems in all areas. These network technologies become increasingly more complex in a wide sphere. The mainstream of software development environment is the development paradigms such as concurrent distributed development environment and so-called open source project by using network computing technologies. Especially, an OSS (open source software) system is frequently applied as server use, instead of client use. Such OSS systems which serve as key components of critical infrastructures in our society are still ever-expanding now. The open source project contains special features of so-called software composition by which several geographically-dispersed components are developed in all parts of the world. The successful experience of adopting the distributed development paradigms in such open source projects includes GNU/Linux operating system[a], Apache Web server, and so on[b] [1]. However, the poor handling of quality attainment and customer support prohibit the progress of OSS. We focus on the problems in the software quality that prohibit the progress of OSS.

Especially, software reliability growth models (SRGM's)[2] have been applied to assess the reliability for quality management and testing-progress control of software development. On the other hand, the effective method of dynamic testing management for new distributed development paradigm as typified by the open source project has only a few presented[3,4]. In case of considering the effect of the debugging process on entire system in the development of a method of reliability assessment for OSS, it is necessary to grasp the situation of registration for bug tracking system, the degree of maturity of OSS, and so on.

In this paper, we focus on an OSS developed under the open source project. We discuss a useful software reliability assessment method in an open source project as a typical case of new distributed development paradigm. Especially, we propose a software reliability growth model based on stochastic differential equations in order to consider the active state of the open source project. Considering the operation environment of OSS, it is different from conventional software systems developed under the identical organization. Then, the expected number of detected faults continues to increase from the effect of the interaction among various operational environments, i.e., the expected number of detected faults cannot converge to a finite value. Also, in most cases, the detected faults of OSS are not reported

---

[a]Linux is a Registered Trademark of Linus Torvalds.
[b]Other company, product, or service names may be trademarks or service marks of others.

to the bug tracking system at the same time as fault-detection but rather reported to the bug tracking system with the time lag of fault-detection and reporting. Therefore, we assume that the software failure intensity depends on the time, and the software fault-reporting phenomena on the bug tracking system keep an irregular state. Moreover, we assume that the software failure intensity depends on the fault importance levels. Also, we analyze actual software fault-count data to show numerical examples of software reliability assessment for the OSS. We show that our proposed model can assist improvement of quality for OSS systems developed under the open source project.

2. Stochastic Differential Equation Modeling

Let $S(t)$ be the cumulative number of faults detected in the OSS system at operational time t $(t \geq 0)$. Suppose that $S(t)$ takes on continuous real values. Since latent faults in the OSS system are detected and eliminated during the operational phase, $S(t)$ gradually increases as the operational procedures go on. Considering the characteristics of open source software development, the OSS developers report several related-faults when the OSS developers confirm the specific faults of bug tracking system, i.e., OSS developers can be OSS users. Therefore, we assume that the increasing rate of $S(t)$ is proportional to the value $S(t)$ itself. Thus, under common assumptions for software reliability growth modeling, we consider the following linear differential equation:

$$\frac{dS(t)}{dt} = \lambda(t)S(t), \tag{1}$$

where $\lambda(t)$ is the intensity of inherent software failures at operational time t and is a non-negative function.

In most cases, the detected faults of OSS are not reported to the bug tracking system at the same time as fault detect but rather reported to the bug tracking system with the time lag of fault-detection and-reporting. As for the fault-reporting to the bug tracking system, we consider that the software fault-reporting phenomena on the bug tracking system keep an irregular state. Moreover, the addition and deletion of software component is repeated under the development of OSS, i.e., we consider that the software failure intensity depends on the time.

Therefore, we suppose that $\lambda(t)$ in Eq. (1) has the irregular fluctuation. That is, we extend Eq. (1) to the following stochastic differential equation[5,6]:

$$\frac{dS(t)}{dt} = \{\lambda(t) + \sigma\gamma(t)\} S(t), \qquad (2)$$

where σ is a positive constant representing a magnitude of the irregular fluctuation and $\gamma(t)$ a standardized Gaussian white noise.

We extend Eq. (2) to the following stochastic differential equation of an Itô type:

$$dS(t) = \{\lambda(t) + \frac{1}{2}\sigma^2\}S(t)dt + \sigma S(t)dW(t), \qquad (3)$$

where $W(t)$ is a one-dimensional Wiener process which is formally defined as an integration of the white noise $\gamma(t)$ with respect to time t. The Wiener process is a Gaussian process and it has the following properties:

$$\Pr[W(0) = 0] = 1, \qquad (4)$$

$$E[W(t)] = 0, \qquad (5)$$

$$E[W(t)W(t')] = \mathrm{Min}[t, t']. \qquad (6)$$

By using Itô's formula[5,6], we can obtain the solution of Eq. (2) under the initial condition $S(0) = v$ as follows[7]:

$$S(t) = v \cdot \exp\left(\int_0^t \lambda(s)ds + \sigma W(t)\right), \qquad (7)$$

where v is the number of detected faults for the previous software version. Using solution process $S(t)$ in Eq. (7), we can derive several software reliability measures.

Moreover, we define the intensity of inherent software failures, $\lambda(t)$, as follows:

$$\int_0^t \lambda(s)ds = \sum_{i=1}^n p_i(1 - \exp[-\alpha_i t]), \qquad (8)$$

where α_i ($i = 1, 2, \cdots, n$) is an acceleration parameter of the intensity of inherent software failures for the i-th fault importance level, p_i ($\sum_{i=1}^n p_i = 1$) the weight parameter for the i-th fault importance level, and n the number of the applied fault importance levels. We can apply the S-shaped growth curve to Eq. (8) depending on the trends of fault importance level.

3. Method of Maximum-Likelihood

In this section, the estimation method of unknown parameters α and σ in Eq. (7) is presented. Let us denote the joint probability distribution function of the process $S(t)$ as

$$P(t_1, y_1; t_2, y_2; \cdots; t_K, y_K)$$
$$\equiv \Pr[S(t_1) \leq y_1, \cdots, S(t_K) \leq y_K | S(t_0) = v], \quad (9)$$

where $S(t)$ is the cumulative number of faults detected up to operational time t $(t \geq 0)$, and denote its density as

$$p(t_1, y_1; t_2, y_2; \cdots; t_K, y_K)$$
$$\equiv \frac{\partial^K P(t_1, y_1; t_2, y_2; \cdots; t_K, y_K)}{\partial y_1 \partial y_2 \cdots \partial y_K}. \quad (10)$$

Since $S(t)$ takes on continuous values, we construct the likelihood function l for the observed data $(t_k, y_k)(k = 1, 2, \cdots, K)$ as follows:

$$l = p(t_1, y_1; t_2, y_2; \cdots; t_K, y_K). \quad (11)$$

For convenience in mathematical manipulations, we use the following logarithmic likelihood function:

$$L = \log l. \quad (12)$$

The maximum-likelihood estimates α_i^* and σ^* are the values making L in Eq. (12) maximize. These can be obtained as the solutions of the following simultaneous likelihood equations[7]:

$$\frac{\partial L}{\partial \alpha_i} = \frac{\partial L}{\partial \sigma} = 0. \quad (13)$$

4. Software Reliability Assessment Measures

4.1. *Expected numbers of detected faults and their variances*

We consider the mean number of faults detected up to operational time t. The density function of $W(t)$ is given by:

$$f(W(t)) = \frac{1}{\sqrt{2\pi t}} \exp\left\{-\frac{W(t)^2}{2t}\right\}. \quad (14)$$

Information on the current number of detected faults in the system is important to estimate the situation of the progress on the operational procedures. Since it is a random variable in our model, its expected value and variance can be useful measures. We can calculate them from Eq. (7) as follows[7]:

$$E[S(t)] = v \cdot \exp\left(\int_0^t \lambda(s)ds + \frac{\sigma^2}{2}t\right), \tag{15}$$

$$\begin{aligned}\operatorname{Var}[S(t)] &= E[\{S(t) - E[S(t)]\}^2] \\ &= v^2 \cdot \exp\left(2\int_0^t \lambda(s)ds + \sigma^2 t\right)\{\exp(\sigma^2 t) - 1\},\end{aligned} \tag{16}$$

where $E[S(t)]$ is the expected number of faults detected up to time t.

From Eq. (15), we can confirm that the number of detected faults cannot converge to a finite value as the following equation:

$$\lim_{t\to\infty} E[S(t)] = \infty. \tag{17}$$

The operating environment of OSS has the characteristics of the susceptible to various operational environments. Therefore, it is different from conventional software systems developed under the identical organization. Then, the expected number of detected faults continues to increase from the effect of the interaction among various operational environments, i.e., the expected number of detected faults cannot converge to a finite value[8,9,10,11,12].

4.2. Mean times between software failures

The instantaneous mean time between software failures (denoted by MTBF$_I$) is useful to measure the property of the frequency of software failure-occurrence.

Instantaneous MTBF is approximately given by

$$MTBF_I(t) = \frac{1}{E[\frac{dS(t)}{dt}]}. \tag{18}$$

Therefore, we have the following instantaneous MTBF:

$$MTBF_I(t) = \frac{1}{v\left(\lambda(t) + \frac{1}{2}\sigma^2\right) \cdot \exp\left(\int_0^t \lambda(s)ds + \frac{\sigma^2}{2}t\right)}. \tag{19}$$

Also, cumulative mean time between software failures (denoted by MTBF$_C$) is approximately given by

$$MTBF_C(t) = \frac{t}{E[S(t)]}. \tag{20}$$

Table 1. Schedule of release candidate versions in Fedora Core 7.

Date	Event
1 February 2007	Test1 Release
29 February 2007	Test2 Release
27 March 2007	Test3 Release
24 April 2007	Test4 Release
31 May 2007	Fedora 7 General Availability

Therefore, we have the following cumulative MTBF:

$$MTBF_C(t) = \frac{t}{v \cdot \exp\left(\int_0^t \lambda(s)ds + \frac{\sigma^2}{2}t\right)}. \qquad (21)$$

4.3. Coefficient of variation

Also, we can derive the following coefficient of variation from Eq. (7):

$$CV(t) \equiv \frac{\sqrt{\text{Var}[S(t)]}}{\text{E}[S(t)]}. \qquad (22)$$

5. Numerical Illustrations

5.1. Data for numerical illustrations

We focus on the Fedora Core Linux[13] which is one of the operating systems developed under an open source project. The Fedora project is made up of many small-size projects. Fedora is a set of projects, sponsored by Red Hat and guided by the Fedora Project Board[c]. These projects are developed by a large community of people who strive to provide and maintain the very best in free, open source software and standards.

The fault-count data used in this paper are collected in the bug tracking system on the website of Fedora project in May 2007. Especially, we focus on the Kernel component of the Fedora Core Linux. The schedule of release candidate versions in Fedora Core 7 is shown in Table 1.

5.2. Reliability assessment results considering fault levels

Figure 1 shows in the number of detected faults for each fault importance level. The number of detected faults for the end of the Test 1 release

---

[c]Fedora is a trademark of Red Hat, Inc. The Fedora Project is not a supported product of Red Hat, Inc.

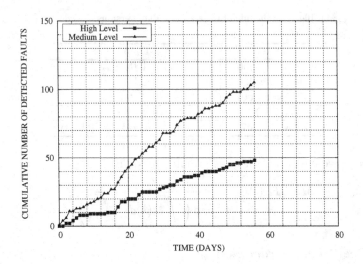

Figure 1. The number of detected faults for each fault level.

version is 48 faults. The number of detected faults for "High" level at the end of fault-reporting on Test 1 is 48 faults, the one of "Medium" level is 105 faults. Therefore, we analyze the actual data about the case of $v = 69$, $p_1 = 0.31373$, and $p_2 = 0.68627$, where v is the number of detected faults before the release of test versions, p_1 the weight of high level, and p_2 the weight of medium level.

The following model parameters have been estimated by solving the likelihood equations given in Eq. (13):

$$\hat{\alpha_1} = 0.035625, \ \hat{\alpha_2} = 0.035509, \ \hat{\sigma} = 0.11385,$$

The estimated expected number of detected faults in Eq. (15), $\widehat{E}[S(t)]$, is shown in Fig. 2.

Also, Figs. 3 and 4 show the estimated software failure intensity, Eq. (8) and $\sum_{i=1}^{n} p_i \alpha_i \exp[-\alpha_i x]$. From Figs. 3 and 4, we can confirm the characteristic of the estimated software failure intensity for each fault importance level.

Figure 5 shows the estimated variance of the number of detected faults in Eq. (16), $\widehat{\text{Var}}[S(t)]$. In Fig. 5, it is shown that the variance of the number of detected faults grows as the time elapses after the evaluation version of Fedora Core 6 has been released.

Moreover, the estimated MTBF$_I$ in Eq. (19), $\widehat{MTBF_I}(t)$, and the estimated MTBF$_C$ in Eq. (21), $\widehat{MTBF_C}(t)$, are also plotted in Figs. 6 and

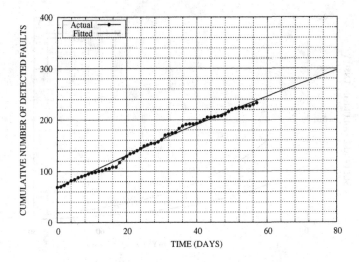

Figure 2. The estimated number of detected faults, $\hat{E}[S(t)]$.

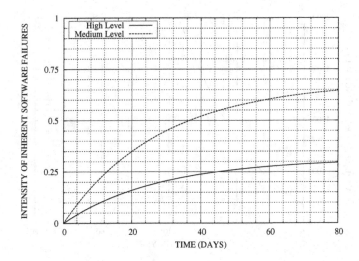

Figure 3. The estimated intensity of inherent software failures for each fault level, $\widehat{S}(t)$.

7, respectively. These figures show that the MTBFs increase as the operational procedures go on. The estimated coefficient of variation, $\widehat{CV}(t)$ is shown in Fig. 8.

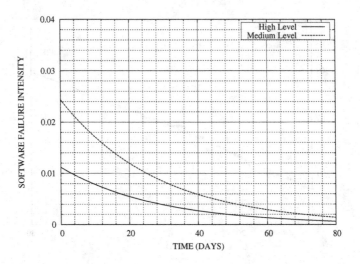

Figure 4. The estimated software failure intensity.

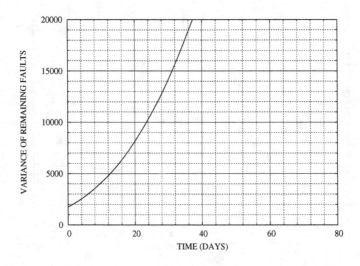

Figure 5. The estimated variance of the number of detected faults, $\widehat{\mathrm{Var}}[S(t)]$.

5.3. Sensitivity analysis in terms of model parameters

From the results of the previous sections, we have verified that our model can be applied to assess quantitatively software reliability in the opera-

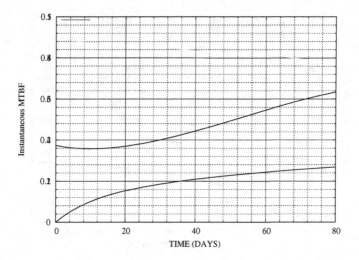

Figure 6. The estimated MTBF$_I$, $\widehat{MTBF}_I(t)$.

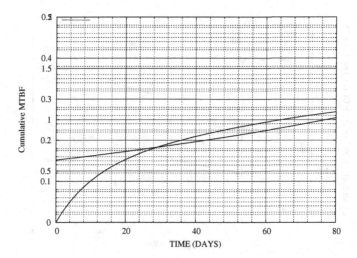

Figure 7. The estimated MTBF$_C$, $\widehat{MTBF}_C(t)$.

tional phase of the OSS. In this section, we show some behavior of software reliability assessment measures if we change parameters σ and α which are the magnitude of the irregular fluctuation and the acceleration parameter of the intensity of initial inherent failure.

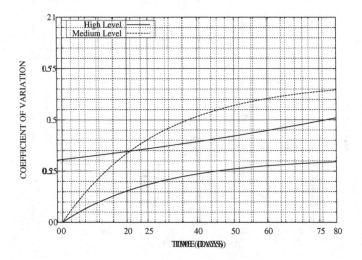

Figure 8. The estimated coefficient of variation, $CV(t)$.

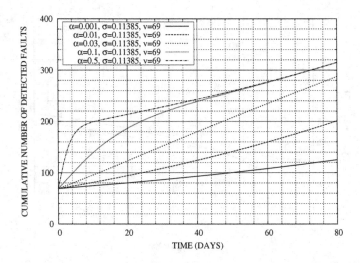

Figure 9. Dependence of model parameter α.

In addition to the case of $(\alpha_1 + \alpha_2)/2 = 0.035567$ and $\sigma = 0.11385$ in the previous section, we represent the estimated mean value function with changing the value of parameters σ and α at regular intervals in Figs. 9 and 10, respectively.

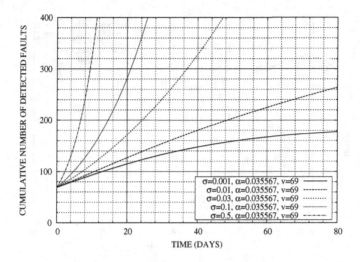

Figure 10. Dependence of model parameter σ.

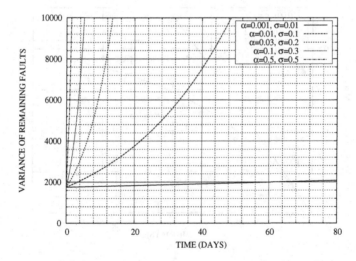

Figure 11. Dependence of model parameter for the variance of the number of detected faults.

Moreover, we show the estimated reliability assessment measures with changing the value of parameters σ and α at regular intervals in Figs. 11–14, respectively.

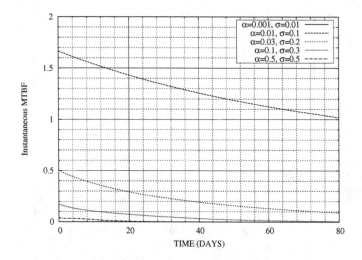

Figure 12. Dependence of model parameter for $MTBF_I$.

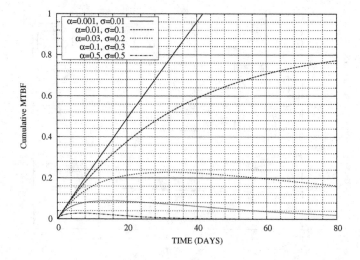

Figure 13. Dependence of model parameter for $MTBF_C$.

6. Concluding Remarks

In this paper, we have focused on the Fedora Core Linux operating system which is well known as the OSS, and discussed the method of reliability assessment for the OSS developed under an open source project.

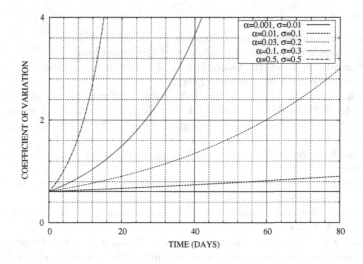

Figure 14. Dependence of model parameter for the coefficient of variation.

Moreover, we have proposed a software reliability growth model based on stochastic differential equations in order to consider the active state of the open source project. Especially, we have assumed that the software failure intensity depends on the time, and the software fault-reporting phenomena on the bug tracking system keep an irregular state. Moreover, we have assumed that the software failure intensity depends on the fault importance levels. Also, we have analyzed actual software fault-count data to show numerical examples of software reliability assessment for the OSS.

Finally, we have focused on an OSS developed under open source projects. New distributed development paradigms typified by such open source project will evolve at a rapid pace in the future. Our proposed method becomes useful as the method of reliability assessment after the release of the evaluation version of OSS, and will assist the improvement of quality for OSS.

In future, we will develop the software reliability assessment tool[14] based on our model and they will be compared with the conventional models based on stochastic differential equations.

Acknowledgements: This work was supported in part by the Grant-in-Aid for Scientific Research (C), Grant No. 18510124 and Young Scientists (B), Grant No. 17700039 from the Ministry of Education, Culture, Sports, Science, and Technology of Japan.

References

1. E-Soft Inc., Internet Research Reports, http://www.securityspace.com/s_survey/data/index.html
2. S. Yamada, *Software Reliability Models: Fundamentals and Applications* (in Japanese), JUSE Press, Tokyo, (1994).
3. A. MacCormack, J. Rusnak, and C.Y. Baldwin, Exploring the structure of complex software designs: an empirical study of open source and proprietary code, *INFORMS Journal of Management Science*, **52** (7), 1015–1030, (2006).
4. G. Kuk, Strategic interaction and knowledge sharing in the KDE developer mailing list, *INFORMS Journal of Management Science*, **52** (7), 1031–1042, (2006).
5. L. Arnold, *Stochastic Differential Equations–Theory and Applications*, John Wiley & Sons, New York, (1974).
6. E. Wong, *Stochastic Processes in Information and Systems*, McGraw–Hill, New York, (1971).
7. S. Yamada, M. Kimura, H. Tanaka, and S. Osaki, Software reliability measurement and assessment with stochastic differential equations, *IEICE Trans. Fundamentals*, **E77-A** (1), 109–116, (1994).
8. Y. Tamura, S. Yamada and M. Kimura, Reliability assessment method based on logarithmic Poisson execution time model for open source project, *Proceedings of the Second IASTED International Multi-Conference on Automation, Control, and Information Technology*, pp. 54–59, (2005).
9. Y. Tamura and S. Yamada, Comparison of software reliability assessment methods for open source software, *Proceedings of the 11th IEEE International Conference on Parallel and Distributed Systems (ICPADS2005)–Volume II*, pp. 488–492, (2005).
10. Y. Tamura and S. Yamada, Validation of an OSS reliability assessment method based on ANP and SRGM's, *Proceedings of the International Workshop on Recent Advances in Stochastic Operations Research*, pp. 273–280, (2005).
11. Y. Tamura and S. Yamada, A method of user-oriented reliability assessment for open source software and its applications, *Proceedings of the 2006 IEEE International Conference on Systems, Man, and Cybernetics*, pp. 2185–2190, (2006).
12. Y. Tamura and S. Yamada, Software reliability assessment and optimal version-upgrade problem for open source software, *Proceedings of the 2007 IEEE International Conference on Systems, Man, and Cybernetics*, pp. 1333–1338, (2007).
13. Fedora Project, sponsored by Red Hat, http://fedora.redhat.com/
14. Y. Tamura, K. Hadatsuki, S. Yamada and M. Kimura, Software tool for estimation of optimal version-upgrade time for open source software (in Japanese), *Proceedings of the Japan Linux Conference 2007*, **5**, http://lc.linux.or.jp/lc2007/, (2007).